HARCOURT

Math

Performance Assessment

Grade 5

Harcourt

Orlando   Austin   Chicago   New York   Toronto   London   San Diego

Visit *The Learning Site!*
www.harcourtschool.com

CONTENTS

Overview .... PA v–PA vi

Administering Performance Assessments .... PA vi

Providing for Students with Special Needs .... PA vi

Scoring Rubrics for Mathematics .... PA vii–PA viii

Blank Rubric Form .... PA ix

Levels of Performance in Harcourt Math .... PA x

Class Record Form .... PA xi–PA xii

Performance Assessment Unit 1

Teacher's Directions .... PA 1
Performance Indicators .... PA 2
Task A   Lunch and a Movie .... PA 3
Task B   Math Game .... PA 4
Annotations .... PA 5
Sample Responses,
     Task A .... PA 6–PA 7
Sample Responses,
     Task B .... PA 8–PA 9

Performance Assessment Unit 2

Teacher's Directions .... PA 10
Performance Indicators .... PA 11
Task A   Ticket Sales .... PA 12
Task B   Histogram Run .... PA 13
Annotations .... PA 14
Sample Responses,
     Task A .... PA 15–PA 16
Sample Responses,
     Task B .... PA 17–PA 18

Performance Assessment Unit 3

Teacher's Directions .... PA 19
Performance Indicators .... PA 20
Task A   At the Stadium .... PA 21
Task B   Jewelry Store .... PA 22
Annotations .... PA 23
Sample Responses,
     Task A .... PA 24–PA 25
Sample Responses,
     Task B .... PA 26–PA 27

▶ Performance Assessment Unit 4

Teacher's Directions .... **PA 28**

Performance Indicators .... **PA 29**

Task A   At the Fair .... **PA 30**

Task B   Rolling Prices .... **PA 31**

Annotations .... **PA 32**

Sample Responses,
    Task A .... **PA 33–PA 34**

Sample Responses,
    Task B .... **PA 35–PA 36**

▶ Performance Assessment Unit 5

Teacher's Directions .... **PA 37**

Performance Indicators .... **PA 38**

Task A   Bake Sale .... **PA 39**

Task B   Clowning Around .... **PA 40**

Annotations .... **PA 41**

Sample Responses,
    Task A .... **PA 42–PA 43**

Sample Responses,
    Task B .... **PA 44–PA 45**

▶   Performance Assessment Unit 6

Teacher's Directions .... **PA 46**

Performance Indicators .... **PA 47**

Task A   Pizza Time .... **PA 48**

Task B   On the Job .... **PA 49**

Annotations .... **PA 50**

Sample Responses,
    Task A .... **PA 51–PA 52**

Sample Responses,
    Task B .... **PA 53–PA 54**

▶ Performance Assessment Unit 7

Teacher's Directions .... **PA 55**

Performance Indicators .... **PA 56**

Task A   The Winning Design .... **PA 57**

Task B   Naming Ordered Pairs .... **PA 58**

Annotations .... **PA 59**

Sample Responses,
    Task A .... **PA 60–PA 61**

Sample Responses,
    Task B .... **PA 62–PA 63**

▶ Performance Assessment Unit 8

Teacher's Directions .... **PA 64**

Performance Indicators .... **PA 65**

Task A   Woodworking .... **PA 66**

Task B   Storage Cabinet .... **PA 67**

Annotations .... **PA 68**

Sample Responses,
    Task A .... **PA 69–PA 70**

Sample Responses,
    Task B .... **PA 71–PA 72**

▶   Performance Assessment Unit 9

Teacher's Directions .... **PA 73**

Performance Indicators .... **PA 74**

Task A   Sale Time .... **PA 75**

Task B   Probability Spins .... **PA 76**

Annotations .... **PA 77**

Sample Responses,
    Task A .... **PA 78–PA 79**

Sample Responses,
    Task B .... **PA 80–PA 81**

Performance Assessment Program

Unique Features of Harcourt Math Performance Assessment

To create assessments that actually evaluate what is taught, it is necessary to target specific math concepts, skills, and strategies at each grade level. In planning the assessment program for *Harcourt Math,* a review was made of performance assessments cited in professional literature and also of those used in state testing programs. Comparisons were made among available models and desirable features were identified. Holistic scoring was chosen as the primary method of scoring. The *Harcourt Math* Performance assessments offer the following features:

- **They model good instruction.**
  The assessments are like mini-lessons.

- **They are diagnostic.**
  By reviewing students' notes, teachers gain valuable insight into the thinking strategies that students are using.

- **They encourage the thinking process.**
  The assessments guide students through the process of organizing their thoughts and revising their strategies as they solve problems.

- **They are flexible.**
  No strict time limits are imposed, and students are encouraged to proceed at their own pace.

- **They use authentic instruction.**
  Each task is based on realistic problem-solving situations.

- **They are scored holistically.**
  Responses are scored by a multi-dimensional rubic to provide a comprehensive view of each student's performance.

Development of the Performance Assessment Program

Each assessment was field-tested with students before it was selected for inclusion in the program. After the assessments were selected, the pool of student papers for each assessment was reviewed and model papers were selected to illustrate the various scores. Annotations were then written for each model paper, explaining why the score was given.

The development process provided an opportunity to drop or correct those assessments that were not working as expected.

Administering the Performance Assessments

- **Be encouraging.**
Your role in administering the assessments should be that of a coach—motivating, guiding, and encouraging students to produce their best work.

- **Be clear.**
The directions for the assessments are not standardized. If necessary, you may rephrase them for students.

- **Be supportive.**
You may assist students who need help. The amount of assistance needed will vary depending on the needs and abilities of your students.

- **Be fair.**
Allow students adequate time to do their best work. They should not feel that they could have done better if they had been given more time.

- **Be flexible.**
All students need not proceed through the assessments at the same rate and in the same manner.

- **Be involving.**
Whenever possible, involve students in the evaluation process.

Providing for Students with Special Needs

Many school districts are facing the challenge of adapting instruction and assessment to make them appropriate for their learners with special needs. Because the performance assessments are not standardized, the procedure for administering them can be adjusted to meet the needs of these learners. Teachers can help students who have difficulty responding by

- pairing a less-proficient learner with a more-proficient learner.

- encouraging students to discuss their ideas with a partner.

- providing an audiotape of the performance assessment and having students read along with the narration.

- permitting students to tape record their responses in place of writing them.

- allowing students to do their initial planning, computing, designing, and drafting on the computer.

- giving students extra time to do their planning.

- providing assistance upon request.

Keep in mind, however, that the more the performance assessments are modified, the less reliable they may be as measures of students' mathematical ability.

Scoring Rubrics for Mathematics

In scoring a student's task, the teacher should ask two questions: *How well did the student use the conventions of mathematics to arrive at a solution?* and *How well did the student communicate the solution?* The scoring system used for the performance assessments is designed to be compatible with those used by many state assessment programs. Using a 4-point scale, the teacher classifies the student's performance as "excellent," "adequate," "limited," or "little or no achievement."

4-Point Scale			
Excellent Achievement	Adequate Achievement	Limited Achievement	Little or No Achievement
3	2	1	0

A **Level 3** paper shows evidence of extensive understanding of content and provides an exceptionally clear and effective solution. A **Level 2** paper shows an acceptable understanding of content and provides a solution that shows reasonable insight. A **Level 1** paper shows partial understanding and is clear in some parts, but not in others. A **Level 0** paper demonstrates poor understanding of content and provides a solution that is unclear.

Sharing Results with Students and Families

The performance assessment can provide valuable insights into students' mathematical abilities by revealing how all students performed on a common task. However, it is important that their performance on the assessment be interpreted in light of other samples that have been collected such as daily papers, student portfolios, and other types of tests, as well as teacher observation.

For Students

Discuss the rubric with students and explain how it will be used. You may even want to score some anonymous papers as a group or have students score each other's papers and discuss the criteria as they apply to those papers. Make photocopies of the rubrics to use for individual reports. Discuss the reports in conferences with students, pointing out their strengths as well as areas in which they could still improve.

For Families

Results of performance assessments may also be shared with families, who will appreciate seeing what their children can do. Show family members the performance assessment so they understand the task the students were asked to perform. Show their child's responses and discuss the strengths and weaknesses of the responses. Explain the scoring rubric and how the responses were evaluated. Show model papers that illustrate the range in student performance to help them put their child's paper in perspective.

Using Results to Assign Grades

No single test, whether a standardized achievement test, a performance assessment, or an open-ended test, can fully measure a student's mathematical ability. For this reason it is important to use multiple measures of assessment. Therefore, a score on performance assessment should not be used as the sole determiner of a report-card grade or semester grade. The performance assessment could represent one of several factors used to determine a student's grade. Assessments could be combined with the results of a selection of tests, daily grades, class participation, self-reflections, and various samples collected in a portfolio. The following table shows how holistic scores can be converted into numerical or letter grades.

Holistic Score	Letter Grade	Numerical Grade
3	A	90-100
2	B	80-89
1	C	70-79
0	D-F	69 or below

Developing Your Own Rubric

A well-written rubric can help teachers score students' work more accurately and fairly. It also gives students a better idea of what qualities their work should exhibit. Using performance assessment to make connections between teaching and learning requires both conceptual and reflective involvement. Determining criteria may be the most difficult aspect of the process of developing assessment criteria on which to evaluate students' performance. Particularly challenging is the task of finding the right language to describe the qualities of student performance that distinguishes mediocre and excellent work. Teachers should begin the process of developing rubrics by

- gathering sample rubrics as models to be adapted as needed.

- selecting samples of students' work that represent a range of quality.

- determining the qualities of work that distinguish good examples from poor examples.

- using those qualities to write descriptors for the desired characteristics.

- continually revising the criteria until the rubric score reflects the quality of work indicated.

To develop your own rubric, you may wish to use a format similar to the models on the next two pages.

Your Own **Scoring Rubric**

Response Level	Performance Indicators
Score 3	**Generally accurate, complete, and clear** __________ __________ __________ __________
Score 2	**Partially accurate, complete, and clear** __________ __________ __________ __________
Score 1	**Minimally accurate, complete, and clear** __________ __________ __________ __________
Score 0	**Not accurate, complete, and clear** __________ __________ __________ __________

Harcourt Math **Scoring Rubric**

Response Level	Levels of Performance
Score 3	**Generally accurate, complete, and clear** ______ All or most parts of the task are successfully completed; the intents of all parts of the task are addressed with appropriate strategies and procedures. ______ There is evidence that the student has a clear understanding of key concepts and procedures. ______ Student work and explanations are clear. ______ Additional illustrations or information, if present, enhance communication. ______ Answers for all parts are correct or reasonable.
Score 2	**Partially accurate, complete, and clear** ______ Some parts of the task are successfully completed; other parts are attempted and their intents addressed, but they are not successfully completed. ______ There is evidence that the student has partial understanding of key concepts and procedures. ______ Some student work and explanations are clear, but it is necessary to make inferences to understand the response. ______ Additional illustrations or information, if present, may not enhance communication significantly. ______ Answers for some parts are correct, but partially correct or incorrect for others.
Score 1	**Minimally accurate, complete, and clear** ______ A part (or parts) of the task is (are) addressed with minimal success while other parts are omitted or incorrect. ______ There is minimal or limited evidence that the student understands concepts and procedures. ______ Student work and explanations may be difficult to follow, and it is necessary to fill in the gaps to understand the response. ______ Additional illustrations or information, if present, do not enhance communication and may be irrelevant. ______ Answers to most parts are incorrect.
Score 0	**Not accurate, complete, and clear** ______ No part of the task is completed with any success. ______ There is little, if any, evidence that the student understands key concepts and procedures. ______ Student work and explanations are very difficult to follow and may be incomprehensible. ______ Any additional illustrations, if present, do not enhance communication and are irrelevant. ______ Answers to all parts are incorrect.

Performance Assessment

Class Record Form

School Teacher **NAMES** Date	Unit 1		Unit 2		Unit 3		Unit 4		Unit 5	
	Task A	Task B	Task A	Task B	Task A	Task B	Task A	Task B	Task A	Task B

Performance Assessment

Class Record Form

School / Teacher / NAMES	Date	Unit 6		Unit 7		Unit 8		Unit 9	
		Task A	Task B	Task A	Task B	Task A	Task B	Task A	Task B

TASK A

Lunch and a Movie

Purpose
To assess student understanding of estimation and problem solving

Grouping
Individuals

Time
10–15 minutes

Introduce the Task
Students estimate the cost of lunch and a movie to find one Rachel can afford. Students determine which lunch combination allows Rachel enough money left to buy popcorn. Students find the amount needed for a movie ticket and the most expensive sandwich and dessert.

TASK B

Math Game

Purpose
To assess student understanding of equations

Grouping
Individual or partners

Time
10–15 minutes

Preparation Hints
Review the properties of addition with the students.

Introduce the Task
Students will write word problems to fit given equations. They will solve equations and identify properties of addition.

TASK A

Lunch and a Movie

Performance Indicators	Observations and Rubric Score
_________ Uses estimation to find combinations of a sandwich and dessert which can be bought for a given amount or less. _________ Finds the actual costs of two combinations of items. _________ Finds the amount of money needed to buy the most expensive sandwich-dessert combination and a movie ticket. _________ Shows work and explains how the answers were determined.	3 2 1 0

TASK B

Math Game

Performance Indicators	Observations and Rubric Score
_________ Chooses an appropriate equation. _________ Writes a word problem for the equation, and then solves the equation. _________ Identifies properties of addition. _________ Shows work and explains how the answers were determined.	3 2 1 0

Total Score _________ /6

Name ___________________________

Lunch and a Movie

Rachel has $10.25 to spend on lunch and a movie. The movie ticket costs $5.

a. Rachel wants a sandwich and a dessert for lunch. Estimate to find one sandwich and one dessert she can order and still have enough money left to buy a ticket for the movie. Then find the actual total cost of these items and compare it with your estimate.

Show your work.

b. List two possible lunches Rachel could have and still have enough money left to buy a box of popcorn for $1.50 after she pays for the movie ticket.

____

____

____

c. Rachel wants to save enough money so that next time she can buy the most expensive sandwich and the most expensive dessert on the menu and still have money left for a movie ticket. How much money will she need?

____

Math Game

Carla and Sandy are playing a math game. They take turns choosing an equation card and writing a word problem for the equation. They exchange problems and solve each other's problems. These are the cards they used.

$14 + n = 25$	$m - 8 = 3$
$0 + y = 37$	$15 - b + 2 = 12$
$9 = 14 - t$	$24 = s + 18$
$28 + d = 28$	$4 + 12 = k + 4$

a. The equation on Carla's card has a subtraction sign. Choose an equation that Carla could have chosen, and write a word problem for that equation. Explain how you would solve that equation.

b. The equation on Sandy's card illustrates an addition property. Choose an equation that Sandy could have chosen. Tell which property the equation illustrates. Explain how you know.

Show your work.

Lunch and a Movie

Rachel has $10.25 to spend on lunch and a movie. The movie ticket costs $5.

a. Rachel wants a sandwich and a dessert for lunch. Estimate to find one sandwich and one dessert she can order and still have enough money left to buy a ticket for the movie. Then find the actual total cost of these items and compare it with your estimate.

Rachel can have a cookie and any sandwich, or a yogurt cone and any sandwich except fish, or a fruit cup with a tuna salad or veggie sandwich.

Movie Café Menu

Sandwiches

tuna salad	$2.25
fish	$3.79
hamburger	$3.19
veggie	$1.89

Desserts

cookie	$0.89
fruit cup	$2.65
yogurt cone	$1.79

Show your work.

	Tuna salad	Fish	Hamburger	Veggie
Cookie	3.14	4.68	4.08	2.78
Fruit cup	4.90	6.44	5.84	4.54
Yogurt cone	4.04	5.58	4.98	3.68

b. List two possible lunches Rachel could have and still have enough money left to buy a box of popcorn for $1.50 after she pays for the movie ticket.

Rachel can have a cookie with the tuna salad sandwich, or a cookie with the veggie sandwich and still have $1.50 for popcorn. (A yogurt cone with the veggie sandwich is another possible lunch.)

c. Rachel wants to save enough money so that next time she can buy the most expensive sandwich and the most expensive dessert on the menu and still have money left for a movie ticket. How much money will she need?

$3.79 + $2.65 − $5 = $11.44 needed for the most expensive sandwich and dessert with enough for a movie.

Math Game

Carla and Sandy are playing a math game. They take turns choosing an equation card and writing a word problem for the equation. They exchange problems and solve each other's problems. These are the cards they used.

$14 + n = 25$	$m - 8 = 3$
$0 + y = 37$	$15 - b + 2 = 12$
$9 = 14 - t$	$24 = s + 18$
$28 + d = 28$	$4 + 12 = k + 4$

a. The equation on Carla's card has a subtraction sign. Choose an equation that Carla could have chosen, and write a word problem for that equation. Explain how you would solve that equation.

b. The equation on Sandy's card illustrates an addition property. Choose an equation that Sandy could have chosen. Tell which property the equation illustrates. Explain how you know.

Show your work.

a. Check students' word problems and explanations.

b. Possible answer: $28 + d = 28$; Identify Property of Addition; When zero is added to a number, the sum is that number.

Lunch and a Movie

Name _______________

Lunch and a Movie

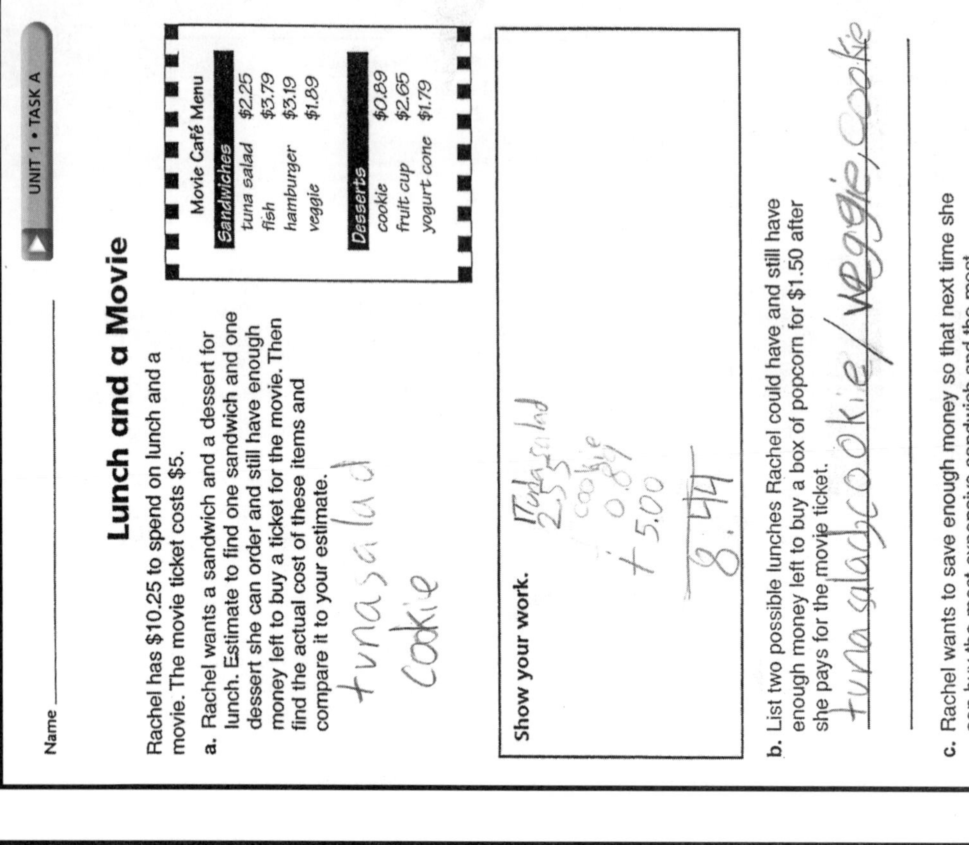

Rachel has $10.25 to spend on lunch and a movie. The movie ticket costs $5.

a. Rachel wants a sandwich and a dessert for lunch. Estimate to find one sandwich and one dessert she can order and still have enough money left to buy a ticket for the movie. Then find the actual cost of these items and compare it to your estimate.

Show your work.

b. List two possible lunches Rachel could have and still have enough money left to buy a box of popcorn for $1.50 after she pays for the movie ticket.

c. Rachel wants to save enough money so that next time she can buy the most expensive sandwich and the most expensive dessert on the menu and still have money left for a movie ticket. How much money will she need?

Level 2 In part *a* this student shows correct computation but does not estimate. Part *b* is correct but no work is shown. Part *c* is correct but expression is not shown.

Name _______________

Lunch and a Movie

Rachel has $10.25 to spend on lunch and a movie. The movie ticket costs $5.

a. Rachel wants a sandwich and a dessert for lunch. Estimate to find one sandwich and one dessert she can order and still have enough money left to buy a ticket for the movie. Then find the actual cost of these items and compare it to your estimate.

Show your work.

b. List two possible lunches Rachel could have and still have enough money left to buy a box of popcorn for $1.50 after she pays for the movie ticket.

c. Rachel wants to save enough money so that next time she can buy the most expensive sandwich and the most expensive dessert on the menu and still have money left for a movie ticket. How much money will she need?

Level 3 The student shows understanding of the task. All steps are addressed. Answers are correct, and all work is shown.

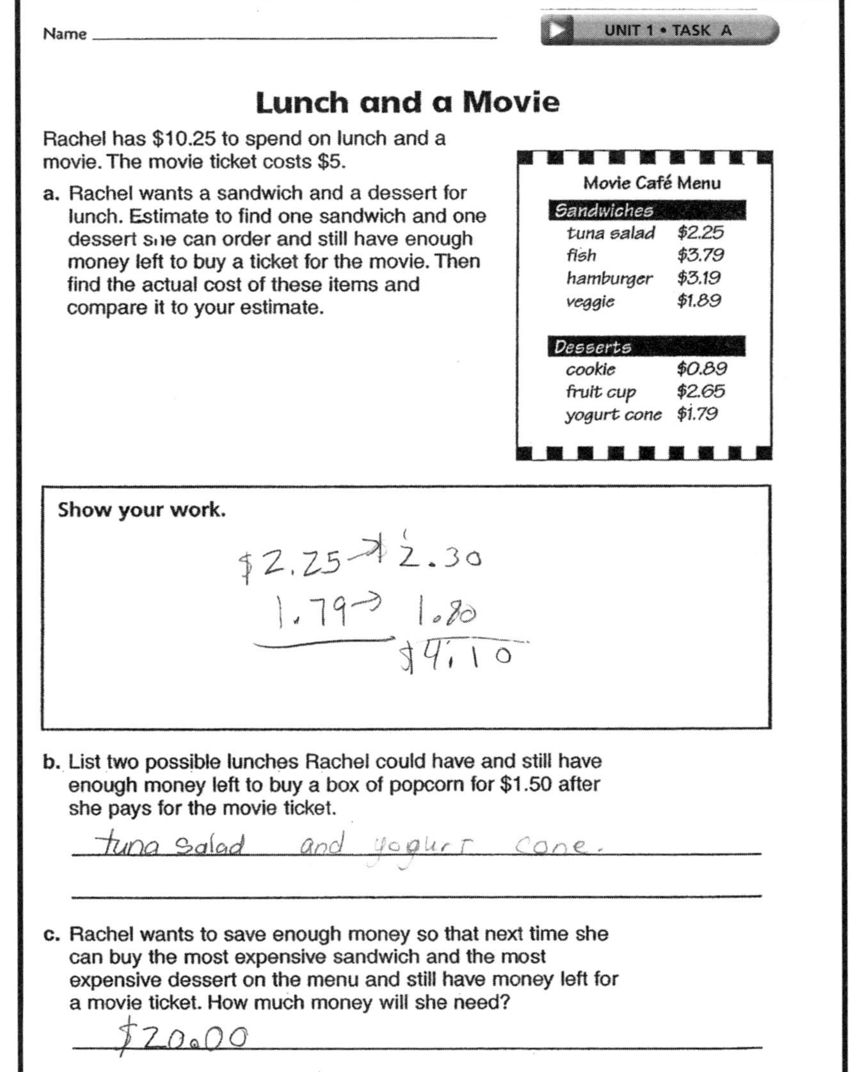

Level 1   Lack of understanding is displayed. Student gives appropriate
sandwich and dessert in part *a*, but parts *b* and *c* are incorrect.

Math Game

Name ___________________________

UNIT 1 • TASK B

Carla and Sandy are playing a math game. They take turns choosing an equation card and writing a word problem for the equation. They exchange problems and solve each other's problems. These are the cards they used.

$14 + n = 25$	$m - 8 = 3$
$0 + y = 37$	$15 - b + 2 = 12$
$9 = 14 - t$	$24 = s + 18$
$28 + d = 28$	$4 + 12 = k + 4$

a. The equation on Carla's card has a subtraction sign. Choose an equation that Carla could have chosen, and write a word problem for that equation. Explain how you would solve that equation.

b. The equation on Sandy's card illustrates an addition property. Choose an equation that Sandy could have chosen. Tell which property the equation illustrates. Explain how you know.

Show your work.

A. Rachel gave away 8 pencils to all her friends. She has 3 left. how many pencils did Rachel begin with.

brainstormed to find that $11 - 8 = 3$.

b. $28 + d = 28$

Zero doesn't change the number

identity property of addition

Level 3 This student shows a good understanding of the task. The student selects an equation, writes a word problem for it, and explains how to solve it, and selects an equation that represents the Identity Property of Addition.

Math Game

Name ___________________________

UNIT 1 • TASK B

Carla and Sandy are playing a math game. They take turns choosing an equation card and writing a word problem for the equation. They exchange problems and solve each other's problems. These are the cards they used.

$14 + n = 25$	$m - 8 = 3$
$0 + y = 37$	$15 - b + 2 = 12$
$9 = 14 - t$	$24 = s + 18$
$28 + d = 28$	$4 + 12 = k + 4$

a. The equation on Carla's card has a subtraction sign. Choose an equation that Carla could have chosen, and write a word problem for that equation. Explain how you would solve that equation.

b. The equation on Sandy's card illustrates an addition property. Choose an equation that Sandy could have chosen. Tell which property the equation illustrates. Explain how you know.

Show your work.

Mack had eleven basketballs he gave his friend Jeff 8 and he have 3 Left.

b $28 + d = 28$ identify property of addiation

$28 + 0 = 28$

Level 2 This work shows partial understanding but is incomplete. No explanations are given for the student's answers to parts *a* and *b*.

Name ___________________________

Math Game

Carla and Sandy are playing a math game. They take turns choosing an equation card and writing a word problem for the equation. They exchange problems and solve each other's problems. These are the cards they used.

$14 + n = 25$	$m - 8 = 3$
$0 + y = 37$	$15 - b + 2 = 12$
$9 = 14 - t$	$24 = s + 18$
$28 + d = 28$	$4 + 12 = k + 4$

a. The equation on Carla's card has a subtraction sign. Choose an equation that Carla could have chosen, and write a word problem for that equation. Explain how you would solve that equation.

b. The equation on Sandy's card illustrates an addition property. Choose an equation that Sandy could have chosen. Tell which property the equation illustrates. Explain how you know.

Show your work.

a. $m - 8 = 3$
Sandy had 11 Balls She gave 8 to carla how much do she have now?

B. $14 + n = 25$
Carla has 14 Puzzles and 11 Jigsaw Puzzles how many do she have?

Level 1   This work is incomplete. The student did not explain how to solve the first word problem. No property was identified for the second equation.

TASK A

Ticket Sales

Purpose
To assess student understanding of descriptive statistics and graphing

Materials
Rulers, grid paper (TR32)

Grouping
Individual or partners

Time
15–20 minutes

Preparation Hints
Describe the different types of graphs and the most appropriate use for each.

Introduce the Task
Students will examine data for ticket sales at two theaters. They will choose and make a graph to best display the data.

TASK B

Histogram Run

Purpose
To assess student understanding of ways to display data

Grouping
Individual or partners

Time
20–30 minutes

Preparation Hints
Discuss histograms and how they differ from other types of graphs.

Introduce the Task
Students use data to make a frequency table. Then students use frequency table to make a histogram.

TASK A

Ticket Sales

Performance Indicators	Observations and Rubric Score
_________ Finds the mean, median, mode, and range for sets of numbers. _________ Chooses the best theater based on the results. _________ Makes a graph to display the data in the table. _________ Shows work and explains how the answers were determined.	3 2 1 0

TASK B

Histogram Run

Performance Indicators	Observations and Rubric Score
_________ Groups data into equal-sized intervals and explains how intervals were chosen. _________ Creates a frequency table of the data using the intervals. _________ Creates a histogram using the frequency table.	3 2 1 0

Total Score _________/6

Ticket Sales

Two local theaters sell advertising space on the back of their tickets.

a. Use the data in the table to find the mean, median, mode, and range of ticket sales for each theater.

b. Compare the ticket sales at the two theaters. Based on this data, which theater would you choose if you want to advertise?

c. Decide what kind of graph would be best to organize and display the data in the table. Make the graph. Write a question that can be answered by using the data in your graph.

Month	Number of Tickets Sold	
	Theater A	Theater B
March	200	300
April	500	50
May	150	600
June	500	500
July	450	650

Show your work.

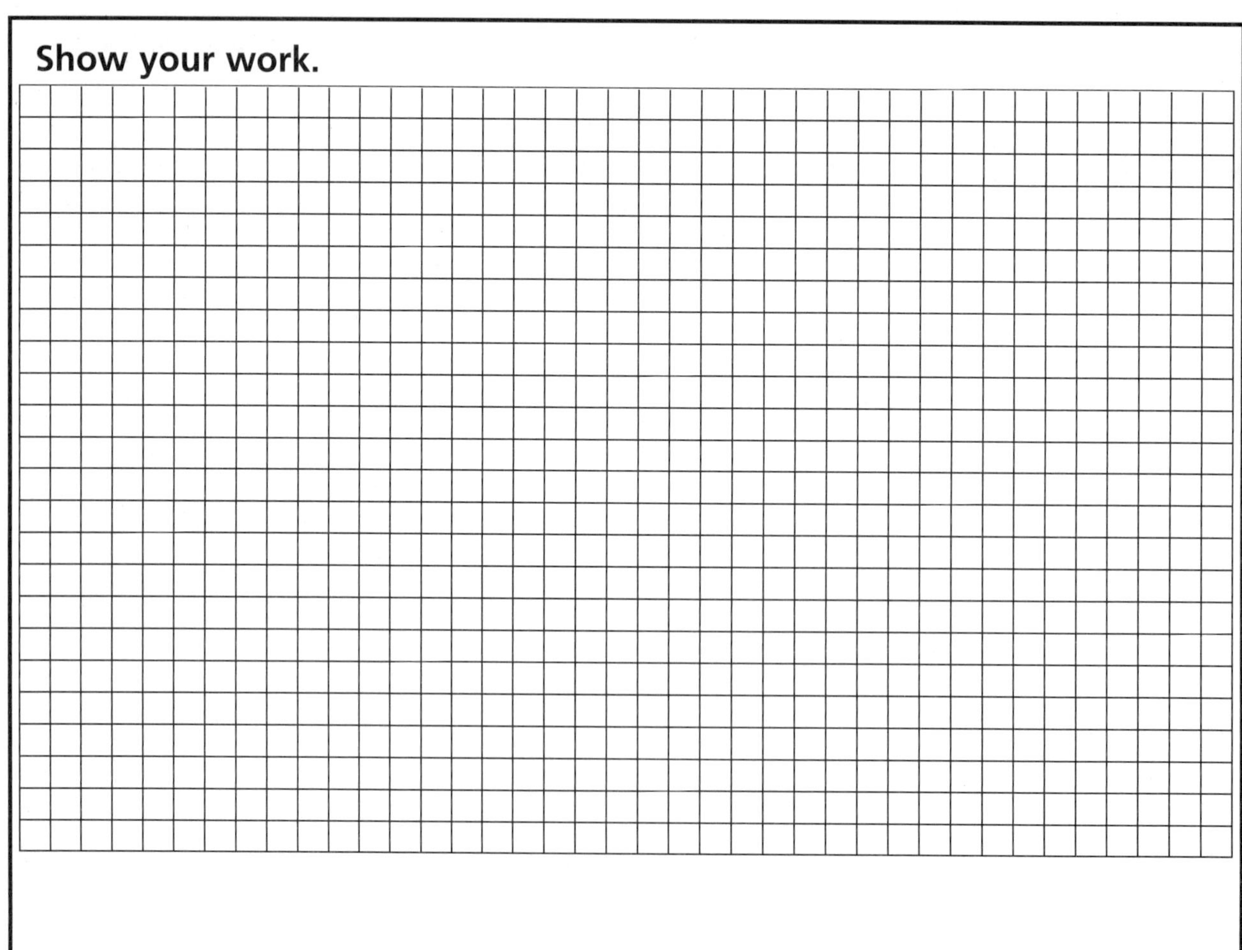

Histogram Run

The table shows the number of runs scored by each player on a Little League team in one season.

RUNS SCORED				
12	15	8	21	32
16	17	15	7	23
24	19	23	9	14

a. Put the data into equal-sized intervals. Explain how you chose the intervals.

b. Use your intervals to make a frequency table of the data.

c. Use your frequency table to make a histogram.

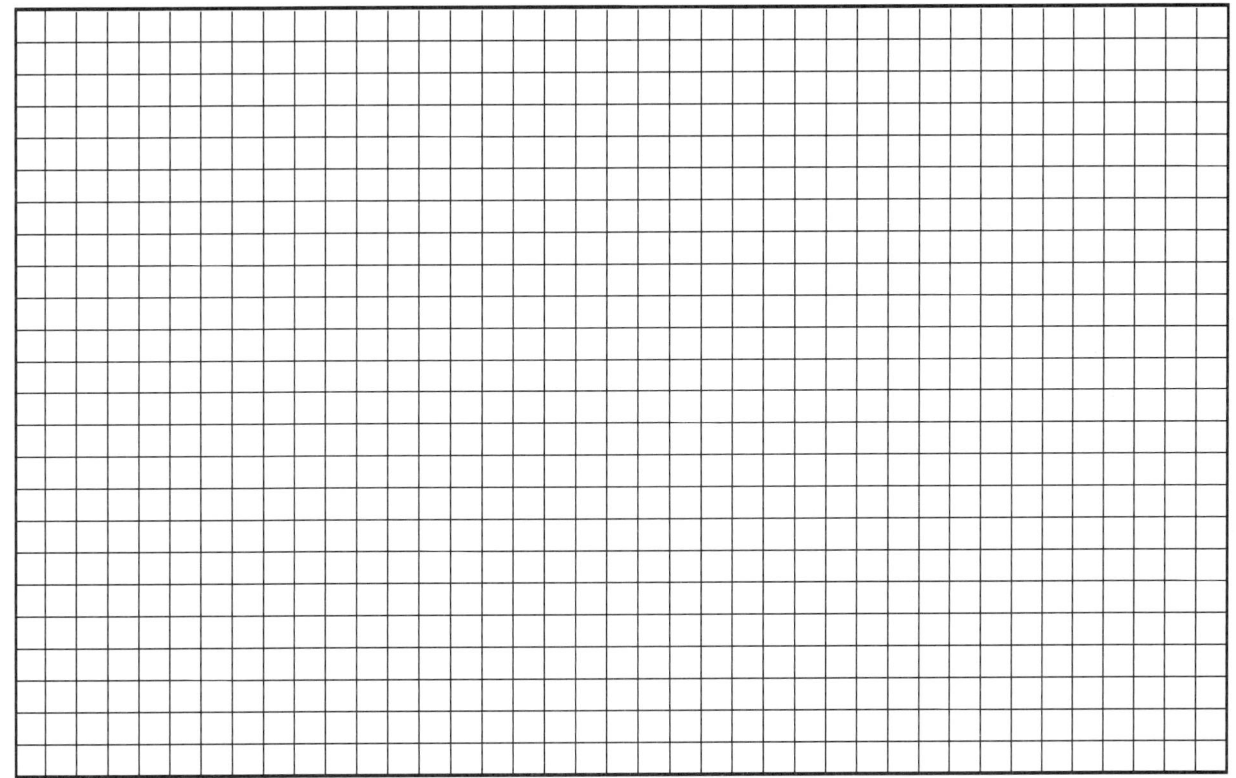

d. Write a question that can be answered using the data in your graph. Then solve. ____

____

Name ___________________________

Ticket Sales

Two local theaters sell advertising space on the back of their tickets.

a. Use the data in the table to find the mean, median, mode, and range of ticket sales for each theater.

b. Compare the ticket sales at the two theaters. Based on this data, which theater would you choose if you want to advertise?

c. Decide what kind of graph would be best to organize and display the data in the table. Make the graph. Write a question that can be answered by using the data in your graph.

Month	Number of Tickets Sold	
	Theater A	Theater B
March	200	300
April	500	50
May	150	600
June	500	500
July	450	650

Show your work.

a. | | A | B |
|---|---|---|
| Mean | 360 | 420 |
| Median | 450 | 500 |
| Mode | 500 | No mode |
| Range | 350 | 600 |

b. Students may pick theater B because it has a greater mean and median number of tickets sold.

c. A double bar graph would best display this data. A double line graph is also a possibility. Sample question that can be answered using the graph: In which month(s) did theater A sell more tickets than theater B?

Name ___________________________

Histogram Run

The table shows the number of runs scored by each player on a Little League team in one season.

RUNS SCORED				
12	15	8	21	32
16	17	15	7	23
24	19	23	9	14

a. Put the data into equal-sized intervals. Explain how you chose the intervals.

b. Use your intervals to make a frequency table of the data.

c. Use your frequency table to make a histogram.

a. Answers may vary. Possible answer: 0–9; 10–19; 20–29; 30–39; because it is easiest to divide the data by the tens digit.

b. Possible answer:

Runs	Frequency
0–9	3
10–19	7
20–29	4
30–39	1

c. Check students' histograms. Possible answer:

d. Write a question that can be answered using the data in your graph. Then solve. Possible answer: How many players scored more than 19 runs?; 5 players

Model Student Papers for
Ticket Sales

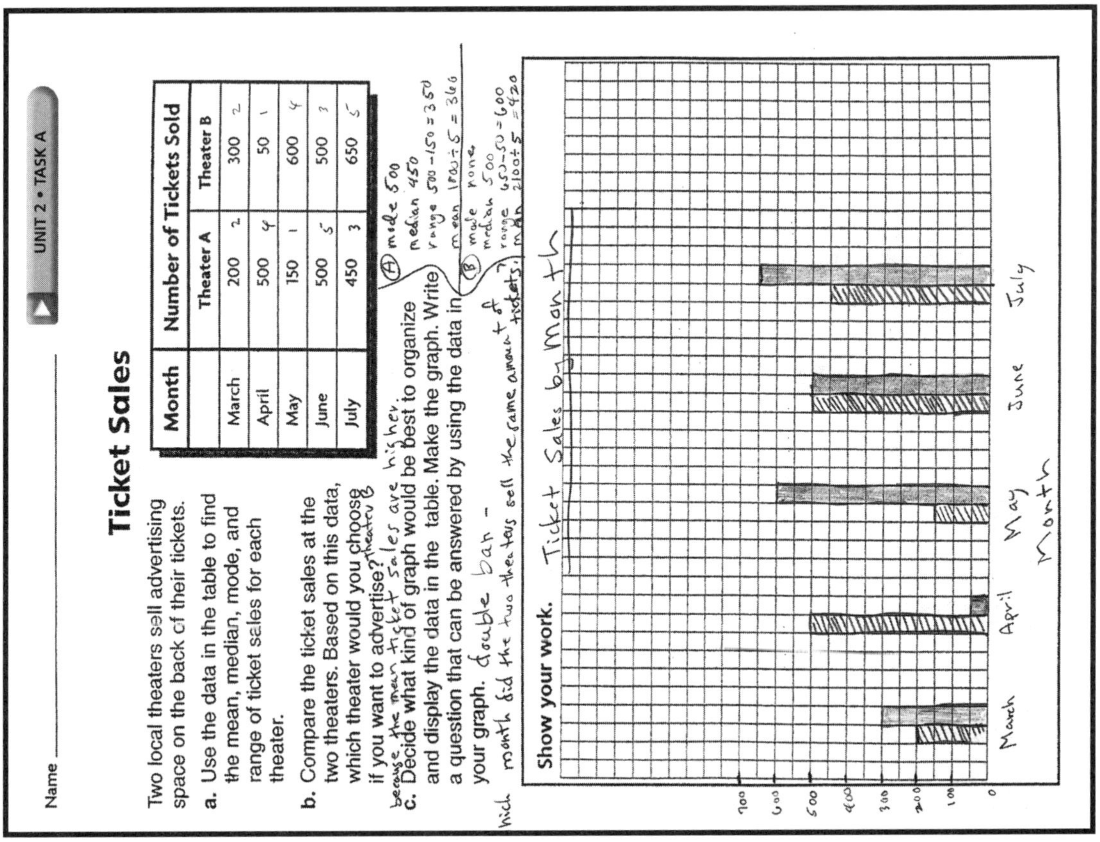

Ticket Sales

Name ______________

Two local theaters sell advertising space on the back of their tickets.

a. Use the data in the table to find the mean, median, mode, and range of ticket sales for each theater.

b. Compare the ticket sales at the two theaters. Based on this data, which theater would you choose if you want to advertise?

c. Decide what kind of graph would be best to organize and display the data in the table. Make the graph. Write a question that can be answered by using the data in your graph.

Month	Number of Tickets Sold	
	Theater A	Theater B
March	200	300
April	500	50
May	150	600
June	500	500
July	450	650

Handwritten annotations (Level 3): Theater A values: 200 2, 500 4, 150 1, 500 5, 450 3. Theater B values: 300 2, 50 1, 600 4, 500 3, 650 5. "because the mean ticket sales are higher. theater B" "double bar –" "hich month did the two theaters sell the same amount of tickets?"

Ⓐ mode 500 / median 450 / range 500 – 150 = 350 / mean 1800 ÷ 5 = 366

Ⓑ mode none / median 500 / range 650 – 50 = 600 / mean 2100 ÷ 5 = 420

Show your work.   Ticket Sales by Month

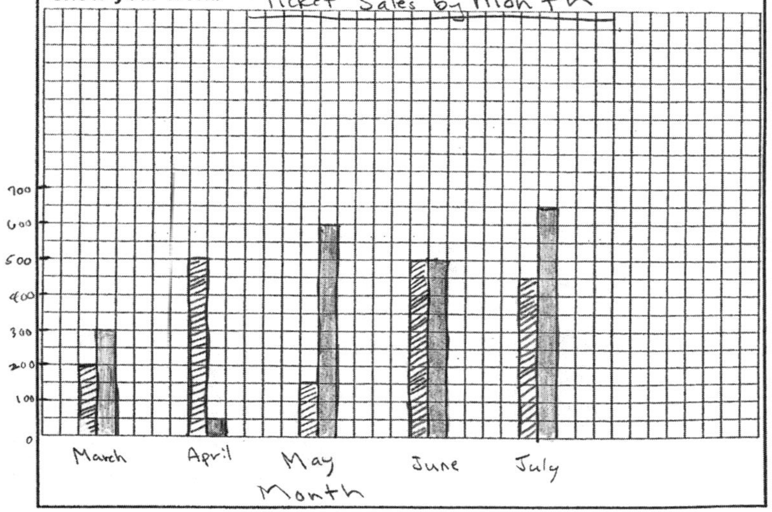

(bar graph with y-axis 0–700, months March, April, May, June, July along x-axis labeled "Month")

Level 3   This student demonstrates a good understanding of the task. The graph is correctly chosen and is correct. All computations are appropriate and correct.

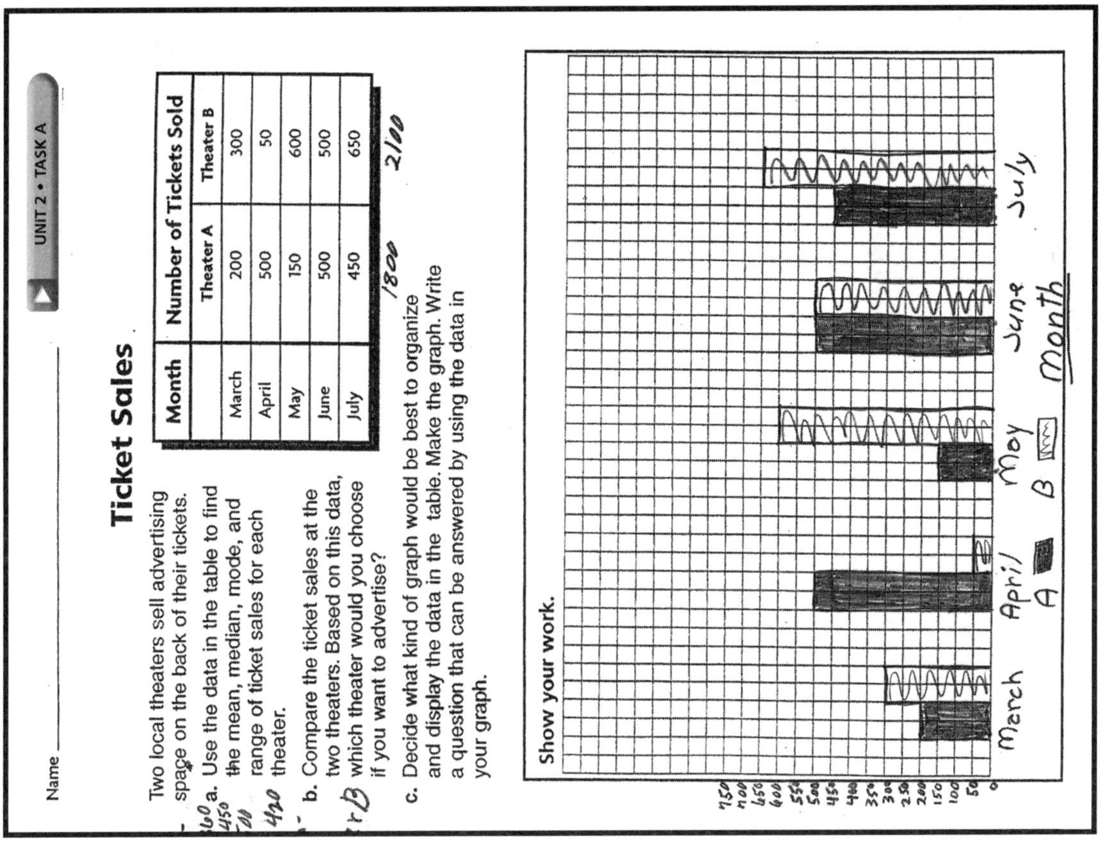

Ticket Sales

Name ______________

Two local theaters sell advertising space on the back of their tickets.

a. Use the data in the table to find the mean, median, mode, and range of ticket sales for each theater.

b. Compare the ticket sales at the two theaters. Based on this data, which theater would you choose if you want to advertise?

c. Decide what kind of graph would be best to organize and display the data in the table. Make the graph. Write a question that can be answered by using the data in your graph.

Month	Number of Tickets Sold	
	Theater A	Theater B
March	200	300
April	500	50
May	150	600
June	500	500
July	450	650

Handwritten annotations (Level 2): "r B" 1800 2100 420

Show your work.

(bar graph with y-axis 0–750 in increments of 50, months March, April, May, June, July along x-axis; A ▭ B ▭ "Month")

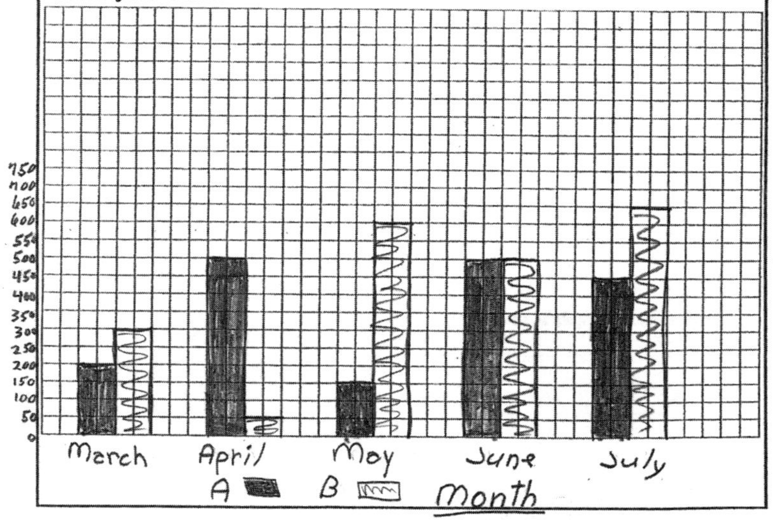

Level 2   This work is correct but incomplete. Graph is appropriate but no title or label for the *y*-axis is given. Student does not formulate a question.

Ticket Sales

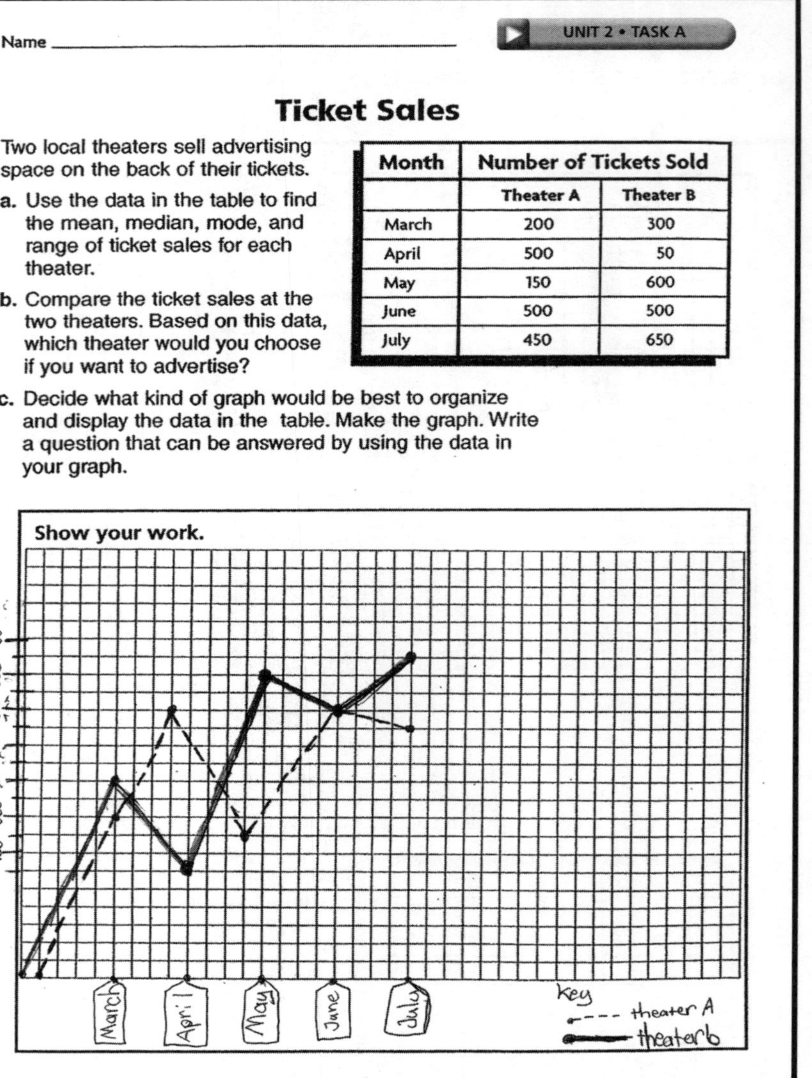

Name _______________________

Ticket Sales

Two local theaters sell advertising space on the back of their tickets.

a. Use the data in the table to find the mean, median, mode, and range of ticket sales for each theater.

b. Compare the ticket sales at the two theaters. Based on this data, which theater would you choose if you want to advertise?

c. Decide what kind of graph would be best to organize and display the data in the table. Make the graph. Write a question that can be answered by using the data in your graph.

Month	Number of Tickets Sold	
	Theater A	Theater B
March	200	300
April	500	50
May	150	600
June	500	500
July	450	650

Level 1 This student shows understanding of graphing by choosing a double line graph, although there is some trouble with the scale on the *y*-axis. Parts *a* and *b* are omitted and no question is posed.

Histogram Run

The table shows the number of runs scored by each player on a little league team in one season.

RUNS SCORED				
12	15	8	21	32
16	17	15	7	23
24	19	23	9	14

a. Put the data into equal-sized intervals. Explain how you chose the intervals.

b. Use your intervals to make a frequency table of the data.

c. Use your frequency table to make a histogram.

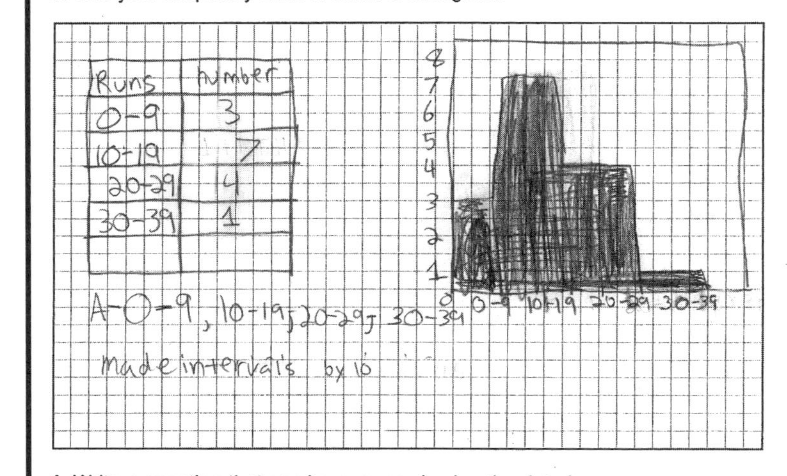

d. Write a question that can be answered using the data in your graph. Then solve. How many people scored less then 10 runs? 3 people.

Level 3   This paper is accurate and complete. The work demonstrates the student's understanding of creating and interpreting intervals, frequency tables, and histograms.

Histogram Run

The table shows the number of runs scored by each player on a little league team in one season.

RUNS SCORED				
12	15	8	21	32
16	17	15	7	23
24	19	23	9	14

a. Put the data into equal-sized intervals. Explain how you chose the intervals.

b. Use your intervals to make a frequency table of the data.

c. Use your frequency table to make a histogram.

d. Write a question that can be answered using the data in your graph. Then solve. What was the largest amont of runs scored?

Level 2   This student grouped the data into intervals and created a frequency table. The student did not explain how the intervals were chosen. The student did make a histogram from the frequency table.

Model Student Papers for
Histogram Run

Name _______________________

Histogram Run

The table shows the number of runs scored by each player on a little league team in one season.

RUNS SCORED				
12	15	8	21	32
16 -	17	15	7	23
24	19	23	9	14

a. Put the data into equal-sized intervals. Explain how you chose the intervals.

b. Use your intervals to make a frequency table of the data.

c. Use your frequency table to make a histogram.

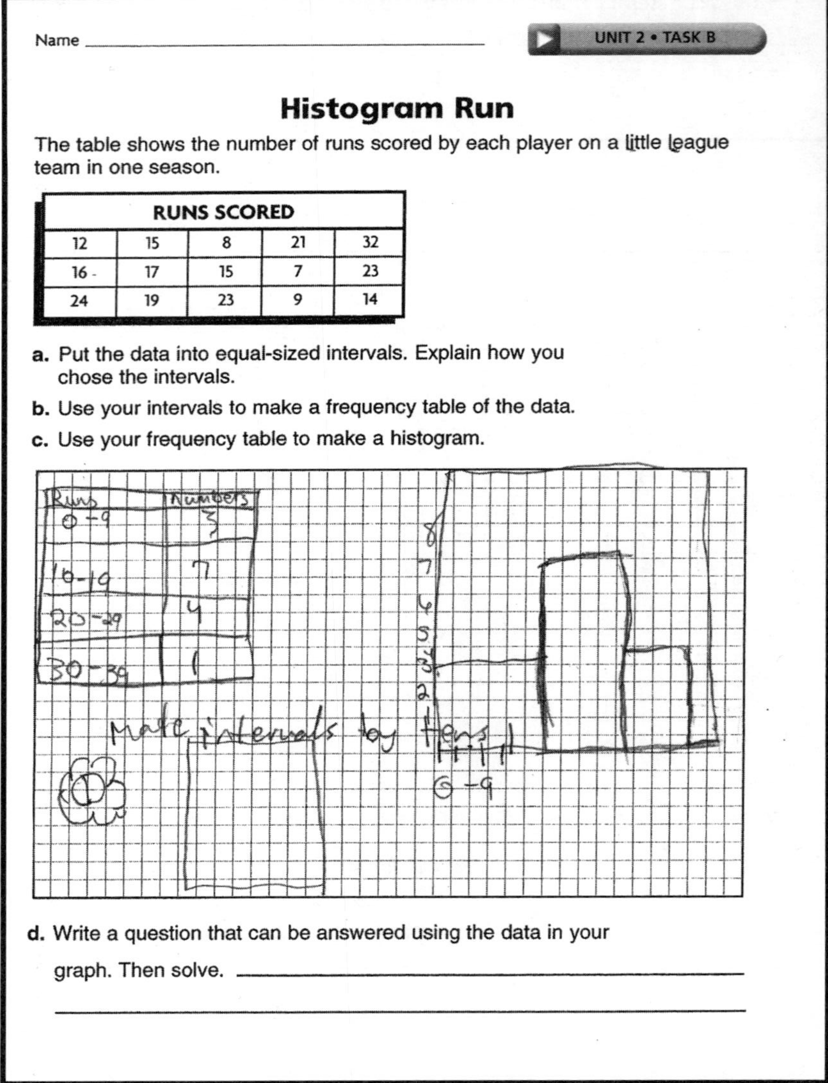

d. Write a question that can be answered using the data in your graph. Then solve. _______________________

____

Level 1   This student displays some understanding of the task, but the work is incomplete.

TASK A

At the Stadium

Purpose
To assess student understanding of estimation and multiplication of whole numbers

Grouping
Individuals or partners

Time
10–15 minutes

Introduce the Task
Students estimate the seating capacity of a stadium. They solve a problem of pricing seats at different amounts to produce a certain amount of money.

TASK B

Jewelry Store

Purpose
To assess student understanding of multiplication of decimals

Grouping
Individuals or partners

Time
10–15 minutes

Preparation Hints
Show students how they can show their design patterns by using initials of colors. For example, they might use YBYBYBYBYBYBYBYB for 16 alternating yellow and blue beads.

Introduce the Task
Students decide on patterns of beads to use for designs of bracelets and necklaces. Students write equations to express and find the length and selling price of their designs.

TASK A

At the Stadium

Performance Indicators	Observations and Rubric Score
_________ Estimates the area of the stadium based on the area of one section. _________ Uses the number of seats in one section to estimate the number of seats in the stadium. _________ Solves a problem of pricing seats in a two-different ticket price plan to produce a given revenue. _________ Shows work and explains how the answers were determined.	3 2 1 0

TASK B

Jewelry Store

Performance Indicators	Observations and Rubric Score
_________ Designs a bracelet with beads and finds its length. _________ Calculates the selling price of the bracelet. _________ Picks a combination of a number of given beads and calculates its length. _________ Shows work and explains how the answers were determined.	3 2 1 0

Total Score _________ /6

At the Stadium

The diagram shows the football stadium at City College.

a. There are 615 seats in section A. What would be a good estimate for the number of seats in section D?

b. At the next football game, 30,000 fans are expected. How can you determine if there will be enough seats?

c. The tickets cost $5 or $10. The stadium operators want to take in at least $200,000 on the sale of tickets for a full stadium. How many seats should sell for $10 and how many for $5?

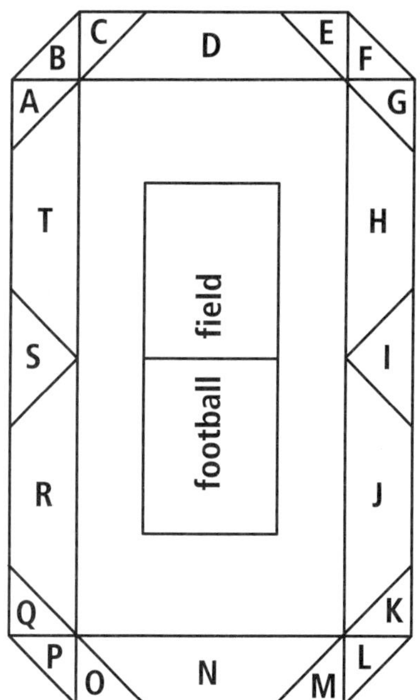

Show your work.

Jewelry Store

Suni uses glass beads to make jewelry. The beads come in six colors. Each color is a different size.

a. Design a bracelet using no fewer than 10 and no more than 20 beads. Include beads of at least two colors. Find the length of the bracelet.

b. Suni sells her bracelets and necklaces for $0.95 per centimeter. How much would the bracelet you designed cost?

c. Suni designs a necklace of 40 beads with red, orange, and yellow colors. What could be the length of the necklace?

Glass Beads	
Color	**Size (in cm)**
red	1.2
blue	0.5
green	0.7
yellow	0.4
orange	1.3
purple	0.6

Show your work.

Name ______________________

At the Stadium

The diagram shows the football stadium at City College.

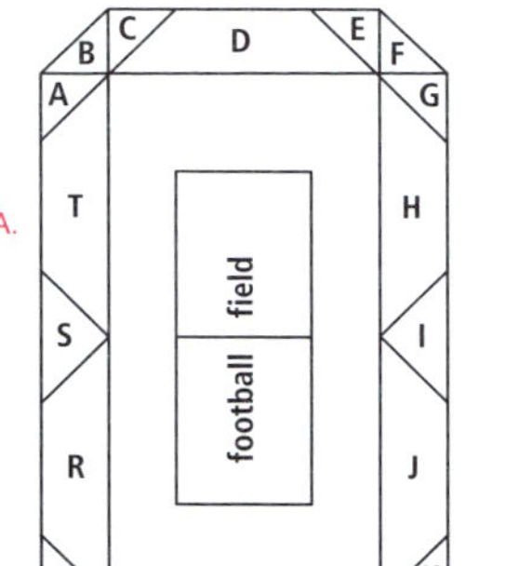

a. There are 615 seats in section A. What would be a good estimate for the number of seats in section D? It looks like section D will hold 6 sections of A. 6 times 615 is about 3,600 seats.

b. At the next football game, 30,000 fans are expected. How can you determine if there will be enough seats? Sections C, D, and E form a rectangle equal to 8 of section A. 8 times 615 is about 5,000 seats.

c. The tickets cost $5 or $10. The stadium operators want to take in at least $200,000 on the sale of tickets for a full stadium. How many seats should sell for $10 and how many for $5?

Show your work.

Including the rectangle formed by C, D, and E, there are 6 rectangles with each holding about 5,000 seats. With 4 "corners" of 615 seats, the stadium can hold more than 30,000 fans.

If all 30,000 seats were sold at $5, the total would be $150,000. If all 30,000 seats were sold at $10, the total would be $300,000. To get $200,000 total, you need $50,000 more than the $150,000 from all seats sold at $5. If 10,000 seats were sold at $10 and the remaining 20,000 were sold at $5, the total would be $200,000.

Note: The total seats are $(6 \times 4,920) + (4 \times 615) = 31,980$.

Name ______________________

Jewelry Store

Suni uses glass beads to make jewelry. The beads come in six colors. Each color is a different size.

a. Design a bracelet using no fewer than 10 and no more than 20 beads. Include beads of at least two colors. Find the length of the bracelet.

b. Suni sells her bracelets and necklaces for $0.95 per centimeter. How much would the bracelet you designed cost?

c. Suni designs a necklace of 40 beads with red, orange, and yellow colors. What could be the length of the necklace?

Glass Beads	
Color	Size (in cm)
red	1.2
blue	0.5
green	0.7
yellow	0.4
orange	1.3
purple	0.6

Show your work.

Sample solution: alternating 10 red and 10 blue beads

$10 \times 1.2 = 12; 10 \times 0.5 = 5$

So the length is 17 centimeters.

The cost is the length times 0.95. For the sample bracelet above,

$17 \times \$0.95 = c$

The cost is $16.15.

Students can divide the 40 beads in any combination of red, orange, and yellow.
Sample solution:
Red   10
Orange   10
Yellow   20
$10 \times 1.2 = 12; 10 \times 1.3 = 13; 20 \times 0.4 = 8$
$12 + 13 + 8$
So, the length could be 33 centimeters.

At the Stadium (Level 3)

UNIT 3 • TASK A

Name ___________

At the Stadium

The diagram shows the football stadium at City College:

a. There are 615 seats in section A. What would be a good estimate for the number of seats in section D?

b. At the next football game, 30,000 fans are expected. How can you determine if there will be enough seats?

c. The tickets cost $5 or $10. The stadium operators want to take in at least $200,000 on the sale of tickets for a full stadium. How many seats should sell for $10 and how many for $5?

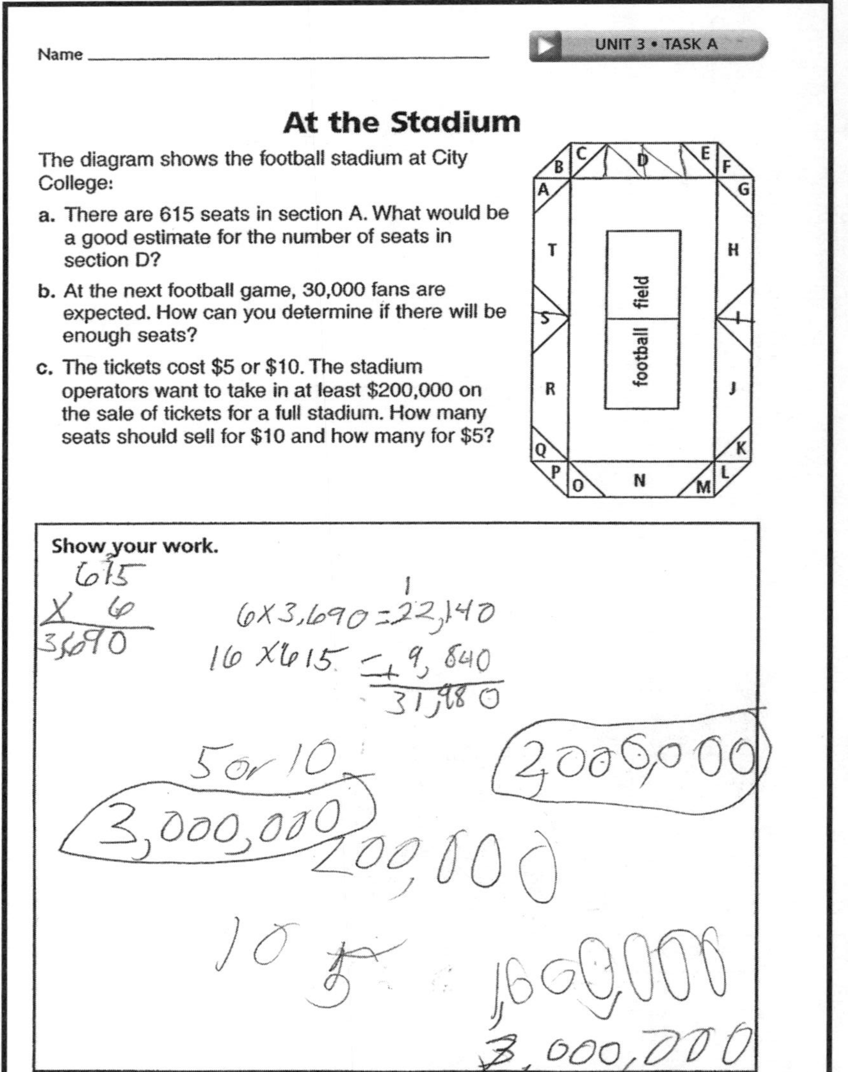

Show your work.

a. Section D
3,690 seats

615
× 6
3690

b. Section D, T, R, N, J, H

3,690
× 6
22,140 seats

Section A, B, C, E, F, G, K, L, M, O, P, Q

615
× 12
1230
615
7,380

Section S, I

615
× 4
2460

22,140
7,380
2,460
31,980 seats

So yes the stadium can hold 30,000 fans.

c. say about 32,000 seats

20,000 × $5 = $100,000
12,000 × $10 = $120,000
$220,000

If they have 20,000 seats for $5 and 12,000 seats for $10 they will be ok.

Level 3 This student shows good understanding of the task. Computations are correct. All tasks are completed.

At the Stadium (Level 2)

UNIT 3 • TASK A

Name ___________

At the Stadium

The diagram shows the football stadium at City College:

a. There are 615 seats in section A. What would be a good estimate for the number of seats in section D?

b. At the next football game, 30,000 fans are expected. How can you determine if there will be enough seats?

c. The tickets cost $5 or $10. The stadium operators want to take in at least $200,000 on the sale of tickets for a full stadium. How many seats should sell for $10 and how many for $5?

Show your work.

615
× 6
3690

6 × 3,690 = 22,140
16 × 615 = 9,840
31,980

5 or 10
3,000,000   2,000,000
200,000

10 5
1,000,000
3,000,000

Level 2 This response is limited. Part *a* is correct. The student gives the correct calculation for part *b*, but the response for part *c* is incomplete and inaccurate.

Name _______________________

At the Stadium

The diagram shows the football stadium at City College:

a. There are 615 seats in section A. What would be a good estimate for the number of seats in section D?

b. At the next football game, 30,000 fans are expected. How can you determine if there will be enough seats?

c. The tickets cost $5 or $10. The stadium operators want to take in at least $200,000 on the sale of tickets for a full stadium. How many seats should sell for $10 and how many for $5?

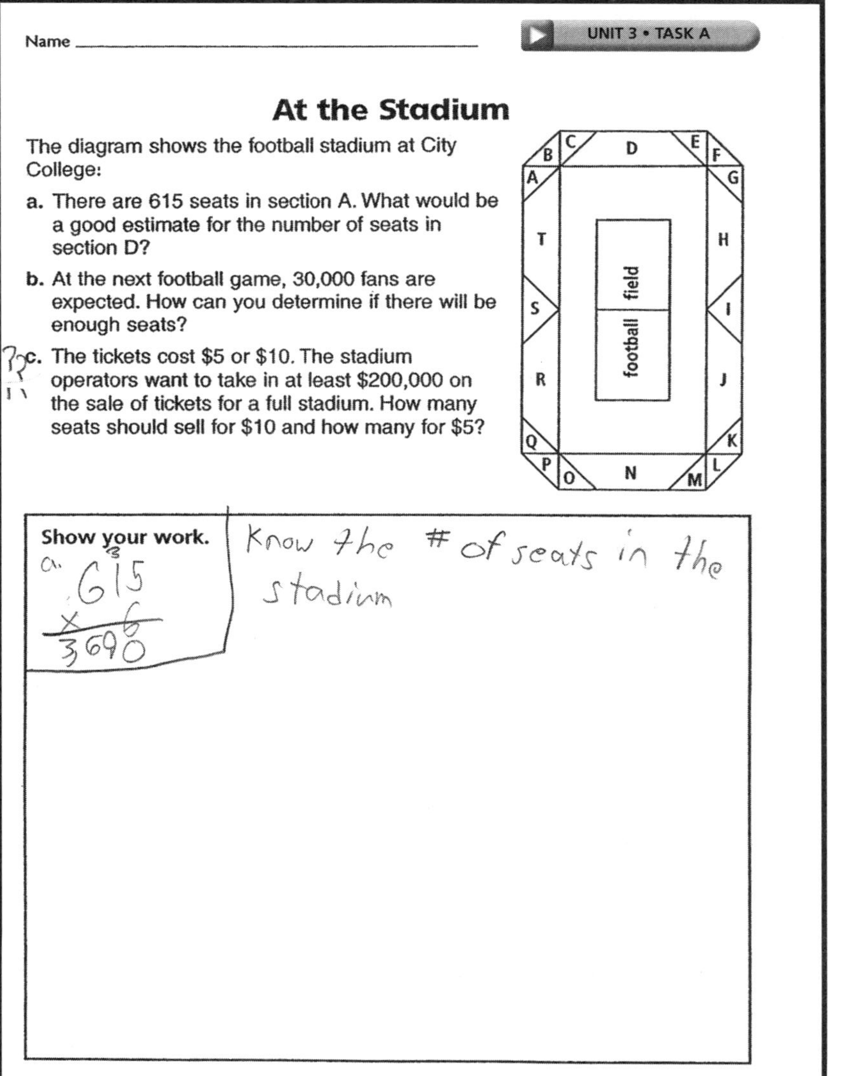

Show your work.

Know the # of seats in the stadium

a. 615
 × 6
 3690

Level 1   Limited understanding is evidenced. Only part *a* is correct. Response to part *b* is incomplete; part *c* is omitted.

Model Student Papers for
Jewelry Store

Left paper (Level 3)

Name ____________________

UNIT 3 • TASK B

Jewelry Store

Suni uses glass beads to make jewelry. The beads come in six colors. Each color is a different size.

a. Design a bracelet using no fewer than 10 and no more than 20 beads. Include beads of at least two colors. Find the length of the bracelet.

b. Suni sells her bracelets and necklaces for $0.95 per centimeter. How much would the bracelet you designed cost?

c. Suni designs a necklace of 40 beads with red, orange, and yellow colors. What could be the length of the necklace?

Glass Beads	
Color	Size (in cm)
red	1.2
blue	0.5
green	0.7
yellow	0.4
orange	1.3
purple	0.6

1.3
× 5
6.5

1.2
× 5
6.0

Show your work.

a. orange – 5 beads   5(1.3) + 5(1.2) + 1(0.5) =
   red – 5 beads   6.5 + 6.0 + 0.5 = 13.00 cm
   blue – 1 bead

b. .95   My bracelet would cost $12.35.
   ×13
   285
   95
   1235

c. yellow – 13 beads .5.2   Suni's necklace
   red – 13 beads 15.6   could be 39.0 cm
   orange – 14 beads 18.2   long.
                    39.0 cm

13   13   14
0.4   1.2   1.3
5.2   26   42
       13   14
       15.6   18.2

Level 3   This student's work demonstrates understanding of the task. The answers are accurate and complete.

Right paper (Level 2)

Name ____________________

UNIT 3 • TASK B

Jewelry Store

Suni uses glass beads to make jewelry. The beads come in six colors. Each color is a different size.

a. Design a bracelet using no fewer than 10 and no more than 20 beads. Include beads of at least two colors. Find the length of the bracelet.

b. Suni sells her bracelets and necklaces for $0.95 per centimeter. How much would the bracelet you designed cost? $5.70

c. Suni designs a necklace of 40 beads with red, orange, and yellow colors. What could be the length of the necklace? It could be 17.7 cm.

4 blues   4 purples
4 yellows
0.5 × 4 = 2
0.6 × 4 = 2.4   0.4 × 4 = 1.6   1.6
                              2.4
                              6.0
.95
× 6
5.70

Glass Beads	
Color	Size (in cm)
red	1.2
blue	0.5
green	0.7
yellow	0.4
orange	1.3
purple	0.6

Show your work.

0.4 × 4 = 1.6   1.6   .95   0.4 × 38 = 15.2
0.5 × 4 = 2   +2.4   × 6
0.6 × 4 = 2.4   6.0   5.70

Level 2   Some of the work shown is correct. In part *a*, the student has correctly computed the length but no equation is written. Part *b* is correct. Part *c* is only partially done.

Jewelry Store

Name ___________________

Jewelry Store

Suni uses glass beads to make jewelry. The beads come in six colors. Each color is a different size.

a. Design a bracelet using no fewer than 10 and no more than 20 beads. Include beads of at least two colors. Find the length of the bracelet.

b. Suni sells her bracelets and necklaces for $0.95 per centimeter. How much would the bracelet you designed cost?

c. Suni designs a necklace of 40 beads with red, orange, and yellow colors. What could be the length of the necklace?

Glass Beads	
Color	Size (in cm)
red	1.2
blue	0.5
green	0.7
yellow	0.4
orange	1.3
purple	0.6

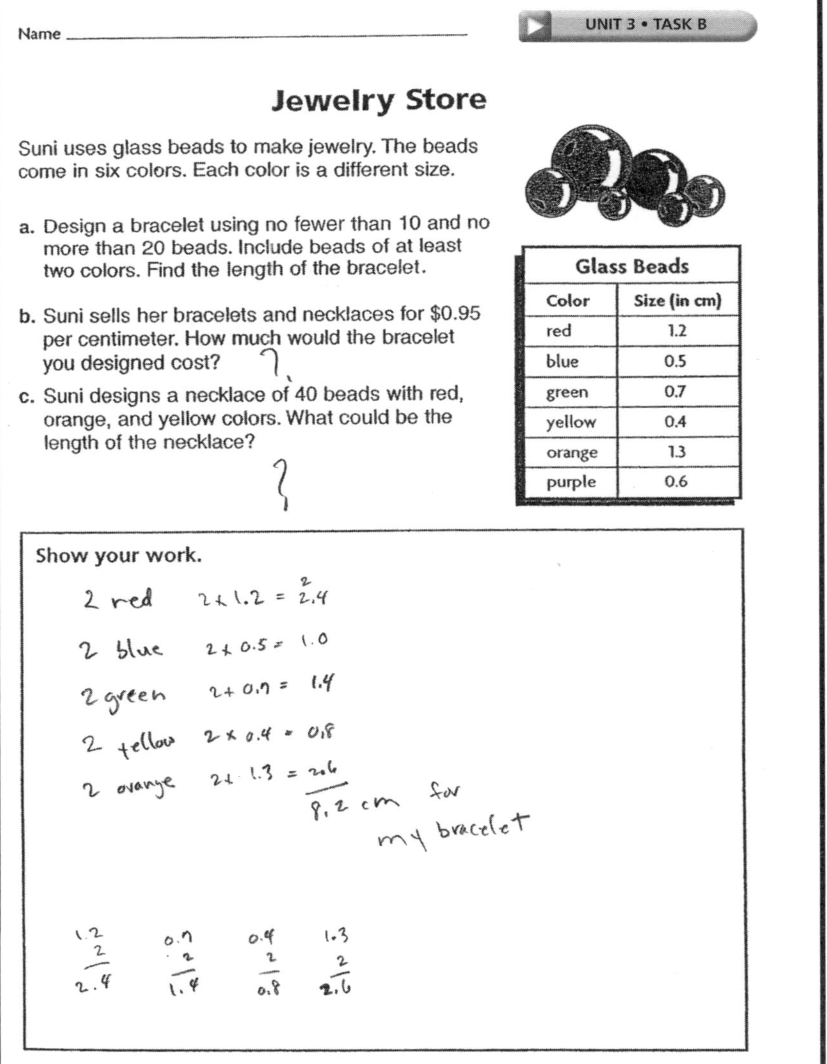

Level 1 Limited understanding is evidenced. Length in c is correctly calculated but no other elements are correct.

TASK A

At the Fair

Purpose
To assess student understanding of using estimates and division of whole numbers to predict and plan

Grouping
Individuals or partners

Time
10–15 minutes

Introduce the Task
Students solve a problem about planning food and seating for a fair. They must estimate, predict, and calculate in their planning.

TASK B

Rolling Prices

Purpose
To assess student understanding of multiplication and division of decimals using money and writing equations

Grouping
Individuals or partners

Time
10–15 minutes

Introduce the Task
Students solve a problem about saving money to make a purchase. Students must calculate how much money must be saved weekly or how many weeks an amount must be saved to buy an item.

TASK A

At the Fair

Performance Indicators	Observations and Rubric Score
_________ Predicts the number of tables needed by estimating and using past information about the fair. _________ Decides how many hamburgers and hot dogs to order by estimating and using past information about the fair. _________ Predicts the number of packages of cheese needed. _________ Shows work and explains how the answers were determined.	3 2 1 0

TASK B

Rolling Prices

Performance Indicators	Observations and Rubric Score
_________ Decides how much Uma should save weekly to buy the scooter. Explain how many weeks Uma will need to save that amount. _________ Determines the number of weeks Jeremy must save in order to purchase the skateboard. Writes an equation using one variable. _________ Determines how much more Thad must save each week to buy the in-line skates than buy the scooter and explains the answer.	3 2 1 0

Total Score _________/6

At the Fair

You are in charge of renting tables and buying hot dogs and hamburgers to sell at the Fall Fair. You want to have enough food, but not much over. Use the notes from last year's fair committee to help you plan.

a. Estimate how many people will eat at the fair. You need to seat about half of the people at one time. How many tables should you rent?

b. How many packages of hot dogs and hamburgers should you buy? Explain why you think these are good estimates.

c. This year you decide to order packages of cheese slices so that half of the hamburger orders can have a slice of cheese. Cheese slices come in packages of 24. How many packages of cheese will you need?

Notes from the Fair

- 475 people ate at the fair.
- More people bought hamburgers than hot dogs. Some people bought both.
- Hot dogs: 8 per package
- Hamburgers: 12 per package
- Tables seat 10 people.

Show your work.

Rolling Prices

Uma, Thad, and Jeremy are shopping for gifts at the Sports Palace.

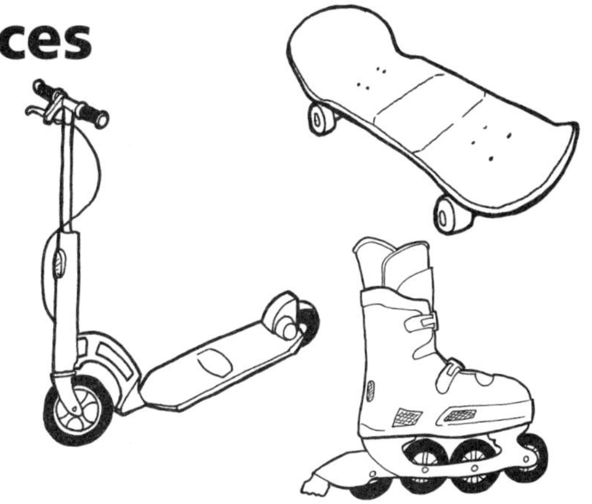

a. Uma earns $10.00 per week. She plans to save the same amount of money each week to buy the scooter for her younger sister. Decide how much Uma should save each week in order to buy the scooter. For how many weeks must she save that amount? Explain.

b. If Jeremy saves $6.00 for *n* number of weeks, how many weeks would he have to save in order to purchase the skateboard? Write an equation using a variable and solve.

Item	Price
Scooter	$39.90
Skateboard	$54.00
In-line Skates	$78.75

c. Thad plans to save the same amount of money each week for 15 weeks to buy the scooter or the in-line skates. How much more must he save each week to buy the in-line skates than to buy the scooter? Explain.

Show your work.

At the Fair

Name ______________________

You are in charge of renting tables and buying hot dogs and hamburgers to sell at the Fall Fair. You want to have enough food, but not much over. Use the notes from last year's fair committee to help you plan.

a. Estimate how many people will eat at the fair. You need to seat about half of the people at one time. How many tables should you rent?

b. How many packages of hot dogs and hamburgers should you buy? Explain why you think these are good estimates.

c. This year you decide to order packages of cheese slices so that half of the hamburger orders can have a slice of cheese. Cheese slices come in packages of 24. How many packages of cheese will you need?

Notes from the Fair

- 475 people ate at the fair.
- More people bought hamburgers than hot dogs. Some people bought both.
- Hot dogs: 8 per package
- Hamburgers: 12 per package
- Tables seat 10 people.

Show your work.

a. About 500 people will eat at the fair. Seating for half the people at one time would require 250 seats. A table seats 10 people so 25 tables are needed.

b. One estimate for food would be:

Hamburgers, 400   50 packages of 8
Hot dogs, 300   25 packages of 12

Students' estimates should have more hamburgers than hot dogs, and more than 500 total since some people bought both.

c. The number of packages of cheese needed will be half the hamburgers divided by 24. For example, for 400 hamburgers, 200 slices would be needed. 200 divided by 24 is about 8. So, about 8 packages of cheese slices would be needed.

Rolling Prices

Name ______________________

Uma, Thad, and Jeremy are shopping for gifts at the Sports Palace.

a. Uma earns $10.00 per week. She plans to save the same amount of money each week to buy the scooter for her younger sister. Decide how much Uma should save each week in order to buy the scooter. For how many weeks must she save that amount? Explain.

b. If Jeremy saves $6.00 for n number of weeks, how many weeks would he have to save in order to purchase the skateboard? Write an equation using a variable and solve.

c. Thad plans to save the same amount of money each week for 15 weeks to buy the scooter or the in-line skates. How much more must he save each week to buy the in-line skates than to buy the scooter? Explain.

Item	Price
Scooter	$39.90
Skateboard	$54.00
In-line Skates	$78.75

Show your work.

a. Possible answer: If Uma saves $5.00 each week for 8 weeks, she will have enough money to buy the scooter.

b. $n \times \$6.00 = \$54.00$; $n = 9$; 9 weeks

c. $2.59 more; for scooter: save $2.66 per week; for in-line skates: save $5.25 per week; $5.25 − $2.66 = $2.59

Left panel (Level 3)

Name ______________________

UNIT 4 • TASK A

At the Fair

You are in charge of renting tables and buying hot dogs and hamburgers to sell at the Fall Fair. You want to have enough food, but not much over. Use the notes from last year's fair committee to help you plan.

a. Estimate how many people will eat at the fair. You need to seat about half of the people at one time. How many tables should you rent? *550 people this year will eat*

b. How many packages of hot dogs and hamburgers should you buy? Explain why you think these are good estimates.
30 packages hamburgers, 40 hotdog packages

c. This year you decide to order packages of cheese slices so that half of the hamburger orders can have a slice of cheese. Cheese slices come in packages of 24. How many packages of cheese will you need?

Notes from the Fair
- 475 people ate at the fair.
- More people bought hamburgers than hot dogs. Some people bought both.
- Hot dogs: 8 per package
- Hamburgers: 12 per package
- Tables seat 10 people.

Show your work. (a) $\frac{1}{2}$ of 550 is 275. To seat around 275 people you should rent about 28 tables.

(b) The Notes say that last year more people bought hamburgers than hotdogs and some bought both.
350 hamburgers + 300 hotdogs = 650 total food
These numbers are big but I think 100 people could buy both foods.
$350 \div 12$ $3 \times 12 = 36$ so you need about 30 packages of hamburgers
$300 \div 8$ $4 \times 8 = 32$ so you need about 40 packages of hotdogs.

(c) $\frac{1}{2}$ of 350 = 175   24 round to 25 and $7 \times 25 = 175$. Since I raised 24 to 25, you better buy 8 packages of cheese slices.

Level 3 The student shows good understanding of the task. The work is correct, all work is shown, all calculations are correct.

Right panel (Level 2)

Name ______________________

UNIT 4 • TASK A

At the Fair

You are in charge of renting tables and buying hot dogs and hamburgers to sell at the Fall Fair. You want to have enough food, but not much over. Use the notes from last year's fair committee to help you plan.

a. Estimate how many people will eat at the fair. You need to seat about half of the people at one time. How many tables should you rent?

b. How many packages of hot dogs and hamburgers should you buy? Explain why you think these are good estimates.

c. This year you decide to order packages of cheese slices so that half of the hamburger orders can have a slice of cheese. Cheese slices come in packages of 24. How many packages of cheese will you need?

Notes from the Fair
- 475 people ate at the fair.
- More people bought hamburgers than hot dogs. Some people bought both.
- Hot dogs: 8 per package
- Hamburgers: 12 per package
- Tables seat 10 people.

C. $24\overline{)150}$ $6\frac{1}{4}$

Show your work.
250 people
A $2\overline{)500}$
150
$2\overline{)300}$ −24 / 10
$10\overline{)250}$ −20 / 50
25 tables
300 hamburgers
+200 hotdogs
500
$8\overline{)200}$ −1 ... 25 pk. hot
$12\overline{)300}$ −24 / 60   25 pk ham. 40

Level 2 Most work is correct. Student did not take into account that some people bought both hamburgers and hot dogs. In part *c* the calculation is correct, but you cannot buy a quarter package of cheese.

Name ________________

At the Fair

You are in charge of renting tables and buying hot dogs and hamburgers to sell at the Fall Fair. You want to have enough food, but not much over. Use the notes from last year's fair committee to help you plan.

a. Estimate how many people will eat at the fair. You need to seat about half of the people at one time. How many tables should you rent?

b. How many packages of hot dogs and hamburgers should you buy? Explain why you think these are good estimates.

c. This year you decide to order packages of cheese slices so that half of the hamburger orders can have a slice of cheese. Cheese slices come in packages of 24. How many packages of cheese will you need?

Notes from the Fair

- 475 people ate at the fair.
- More people bought hamburgers than hot dogs. Some people bought both.
- Hot dogs: 8 per package
- Hamburgers: 12 per package
- Tables seat 10 people.

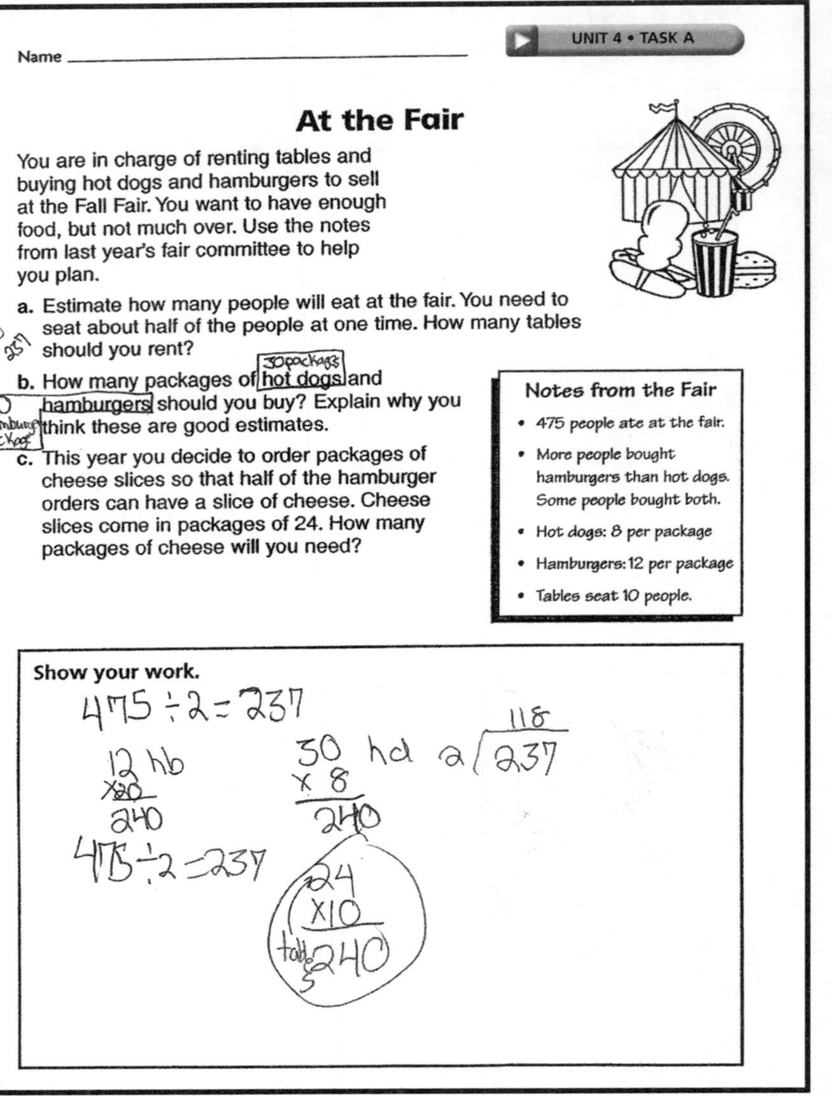

Level 1   The student chooses the same number of attendees as the previous year but number of tables is appropriate. Number of hamburgers and hot dogs is acceptable. Part *c* is omitted.

Model Student Papers for
Rolling Prices

Name _______________

Rolling Prices

Uma, Thad, and Jeremy are shopping for gifts at the Sports Palace.

a. Uma earns $10.00 per week. She plans to save the same amount of money each week to buy the scooter for her younger sister. Decide how much Uma should save each week in order to have enough money to buy the scooter. How many weeks will she need to save that amount? Explain.

b. If Jeremy saves $6.00 for *n* number of weeks, how many weeks would he have to save in order to purchase the skateboard? Write an equation using a variable and solve.

c. Thad plans to save the same amount of money each week for 15 weeks to buy the scooter or the in-line skates. How much more will he have to save each week to buy the in-line skates than to buy the scooter? Explain.

Item	Price
Scooter	$39.90
Skateboard	$54.00
In-line Skates	$78.75

Show your work.

a. Uma should save $8 a week for 5 weeks. This way she will have $40.00   8×5=40

B. Jeremy should save for 9 weeks.
6×n=54
n=9

C. 15×n = 39.90
15×n = 78.75
2.66
15)39.90
5.25
15)78.75
5.25
-2.66
2.59

Theo will have to save $2.59 more each week to buy the skates.

Level 3   This student displays a good understanding of the task. All parts of the task are accurate and complete. The work is shown and explained thoroughly.

Name _______________

Rolling Prices

Uma, Thad, and Jeremy are shopping for gifts at the Sports Palace.

a. Uma earns $10.00 per week. She plans to save the same amount of money each week to buy the scooter for her younger sister. Decide how much Uma should save each week in order to have enough money to buy the scooter. How many weeks will she need to save that amount? Explain.

b. If Jeremy saves $6.00 for *n* number of weeks, how many weeks would he have to save in order to purchase the skateboard? Write an equation using a variable and solve.

c. Thad plans to save the same amount of money each week for 15 weeks to buy the scooter or the in-line skates. How much more will he have to save each week to buy the in-line skates than to buy the scooter? Explain.

Item	Price
Scooter	$39.90
Skateboard	$54.00
In-line Skates	$78.75

Show your work.

A. Uma should save her money for 4 weeks. There are many other ways but this would be the fastest.

B. $6·n=$54   n=9   Jeremy should save for 9 weeks

C. 15·n =$39.90
n=$2.66
2.66
15)39.90
15·n=$78.75
n=$5.25
5.25
15)78.75
$5.25
-$2.66
$2.59

Thad will have to save $2.59 more each week to by the skates not the scooter.

Level 2   This student seems to understand the task. However, part *a* is incomplete. It doesn't specify how much should be saved over four weeks. The other answers are complete and correct.

Rolling Pages

Name _______________________

Rolling Prices

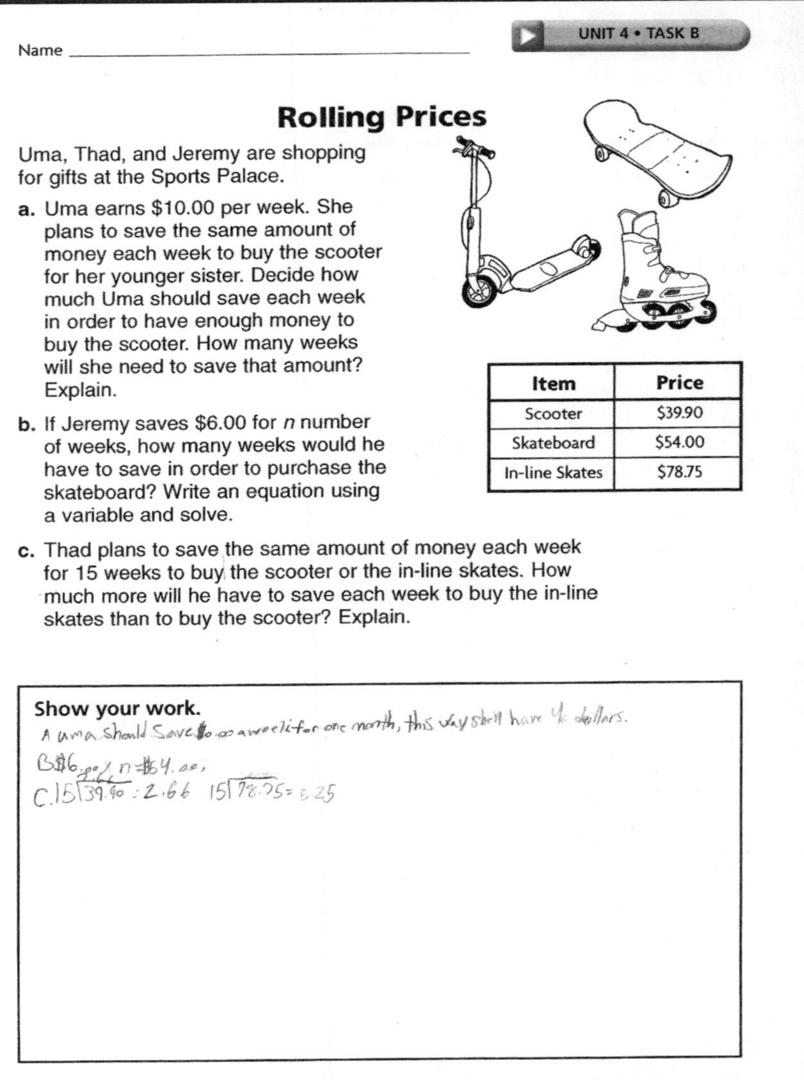

Uma, Thad, and Jeremy are shopping for gifts at the Sports Palace.

a. Uma earns $10.00 per week. She plans to save the same amount of money each week to buy the scooter for her younger sister. Decide how much Uma should save each week in order to have enough money to buy the scooter. How many weeks will she need to save that amount? Explain.

b. If Jeremy saves $6.00 for n number of weeks, how many weeks would he have to save in order to purchase the skateboard? Write an equation using a variable and solve.

c. Thad plans to save the same amount of money each week for 15 weeks to buy the scooter or the in-line skates. How much more will he have to save each week to buy the in-line skates than to buy the scooter? Explain.

Item	Price
Scooter	$39.90
Skateboard	$54.00
In-line Skates	$78.75

Show your work.

A uma should Save $10.00 a week for one month, this way she'll have 40 dollars.

B $6.00 $\times$ n =$54.00,

C. 15)39.90 = 2.66   15)78.75 = 5.25

Level 1 The student shows some understanding, but the solutions are incomplete.

TASK A

Bake Sale

Purpose
To assess student understanding of common factors

Grouping
Individuals or partners

Time
5–10 minutes

Preparation Hints
Point out that each bag will contain all three kinds of cookies.

Introduce the Task
Students solve a problem about how to divide batches of cookies into equal groups. They consider several possible solutions.

TASK B

Clowning Around

Purpose
To assess student understanding of comparison of mixed numbers

Grouping
Individuals or partners

Time
5–10 minutes

Introduce the Task
Students solve a problem about selecting fabric for a clown costume. They must consider the sizes of the pieces of fabric and the amounts needed for different parts of the costume.

Name _______________________________________ Date _______________________

Bake Sale

Performance Indicators	Observations and Rubric Score
_________ Finds the common factors of a set of numbers. _________ Finds the greatest common factor of a set of numbers. _________ Gives multiples of a common factor. _________ Shows work and explains how the answers were determined.	3 2 1 0

Clowning Around

Performance Indicators	Observations and Rubric Score
_________ Selects the least mixed number or fraction. _________ Compares mixed numbers. _________ Solves the problem of finding enough fabric for the pants when no single piece is large enough. _________ Shows work and explains how the answers were determined.	3 2 1 0

Total Score _________ /6

Bake Sale

Ariel, Brad, and Curt baked cookies for the school bake sale.

The students want to put the cookies in bags. Each bag will contain all three kinds of cookies. All the bags will have the same number of each kind of cookie.

Baker	Number of Cookies	Kind of Cookie
Ariel	24	peanut
Brad	18	raisin
Curt	12	oatmeal

a. Find all the ways each kind of cookie can be divided so there is the same number in each bag.

b. What is the greatest number they can put in each bag? How many bags of cookies does this make?

c. Other students want to bring in cookies. Name two other numbers of cookies that can be divided in the same way.

Show your work.

Clowning Around

Jolynn and her mother are making a clown costume for the class play. They found some pieces of fabric at a yard sale. Help them choose fabric pieces for the costume.

a. They can use the shortest piece of fabric for the hat. Which piece should they choose for the hat?

b. They need $2\frac{5}{8}$ yards for the top of the costume. Is there a piece large enough for the top? If so, which is it?

c. Decide what other pieces of fabric they should buy for the sleeves and the pants of the costume. The pants will take more fabric than the top or the sleeves.

Show your work.

Bake Sale

Name ________________

Ariel, Brad, and Curt baked cookies for the school bake sale.

The students want to put the cookies in bags. Each bag will contain all three kinds of cookies. All the bags will have the same number of each kind of cookie.

Baker	Number of Cookies	Kind of Cookie
Ariel	24	peanut
Brad	18	raisin
Curt	12	oatmeal

a. Find all the ways each kind of cookie can be divided so there is the same number in each bag.

b. What is the greatest number they can put in each bag? How many bags of cookies does this make?

c. Other students want to bring in cookies. Name two other numbers of cookies that can be divided in the same way.

Show your work.
a. They can use 2 bags (12 peanut, 9 raisin, 6 oatmeal), 3 bags (8 peanut, 6 raisin, 4 oatmeal), or 6 bags (4 peanut, 3 raisin, 2 oatmeal).
b. The greatest number is 27 cookies in each of 2 bags.
c. Possible answers: 30 cookies and 36 cookies

Clowning Around

Name ________________

Jolynn and her mother are making a clown costume for the class play. They found some pieces of fabric at a yard sale. Help them choose fabric pieces for the costume.

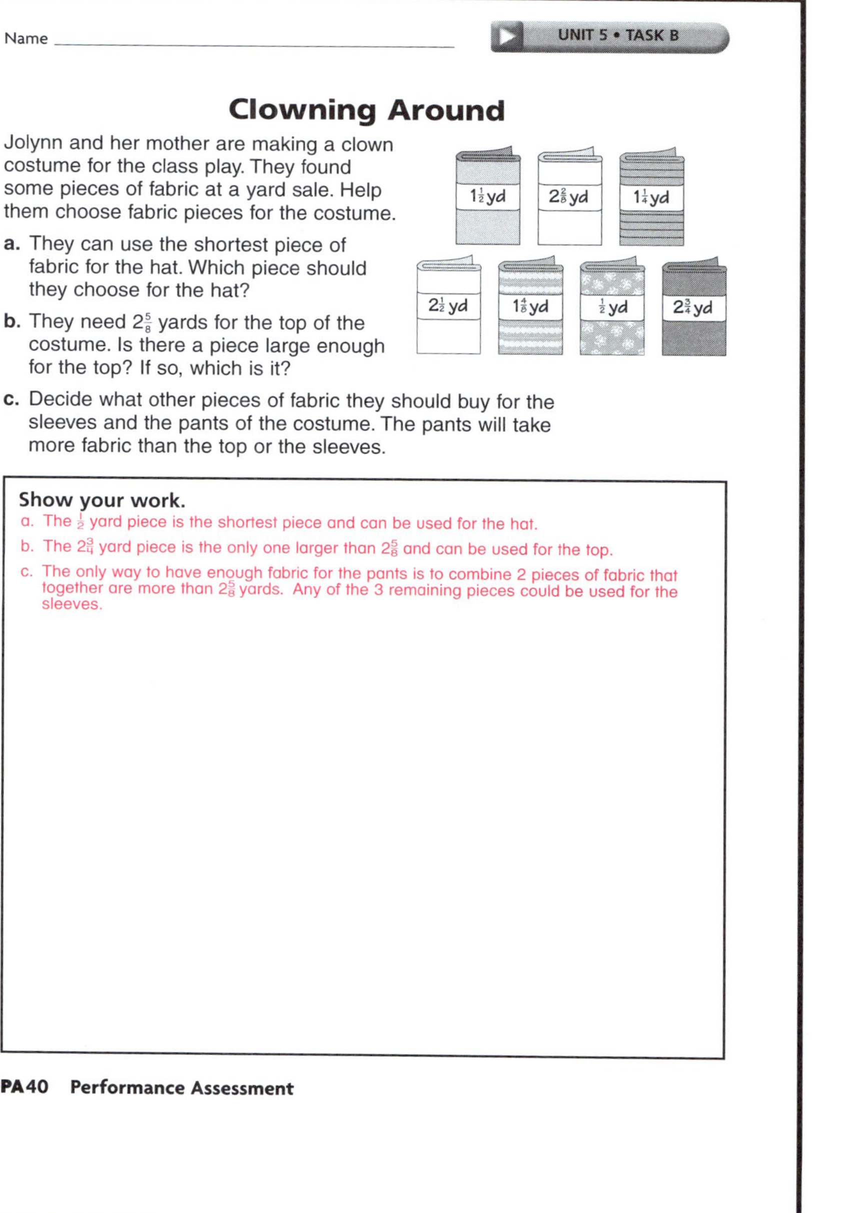

a. They can use the shortest piece of fabric for the hat. Which piece should they choose for the hat?

b. They need $2\frac{5}{8}$ yards for the top of the costume. Is there a piece large enough for the top? If so, which is it?

c. Decide what other pieces of fabric they should buy for the sleeves and the pants of the costume. The pants will take more fabric than the top or the sleeves.

Show your work.
a. The $\frac{1}{2}$ yard piece is the shortest piece and can be used for the hat.
b. The $2\frac{3}{4}$ yard piece is the only one larger than $2\frac{5}{8}$ and can be used for the top.
c. The only way to have enough fabric for the pants is to combine 2 pieces of fabric that together are more than $2\frac{5}{8}$ yards. Any of the 3 remaining pieces could be used for the sleeves.

Name ___________________

UNIT 5 • TASK A

Bake Sale

Ariel, Brad, and Curt baked cookies for the school bake sale.

Baker	Number of Cookies	Kind of Cookie
Ariel	24	peanut
Brad	18	raisin
Curt	12	oatmeal

The students want to put the cookies in bags. Each bag will contain all three kinds of cookies. All the bags will have the same of each kind of cookie.

a. Find all the ways each kind of cookies can be divided so there is the same number in each bag.

b. What is the greatest number they can put in each bag? How many bags of cookies does this make?

c. Other students want to bring in cookies. Name two other numbers of cookies that can be divided in the same way.

Show your work.

Peanut   Raisan   Oatmeal
2x12   2x9   2+6
3x8   3x6   3x4
4x6   6x3   6x2
6x4
8x3

2 bags with 12+9+6=27
3 bags with 8+6+4=18
6 bags with 4+3+2=9

Two bags would have the most, but if they wanted more bags to sell they should use 6 bags.

For 2 bags
30 cookies 2)30 15 in each
10 cookies 2)10 5 in each

For 6 bags
30 cookies 6)30 5 in each
36 cookies 6)36 6 in each

Level 3 This student shows a good understanding of the task. All parts of the task are accurate and complete. All work is shown and is accurate.

Name ___________________

UNIT 5 • TASK A

Bake Sale

Ariel, Brad, and Curt baked cookies for the school bake sale.

Baker	Number of Cookies	Kind of Cookie
Ariel	24	peanut
Brad	18	raisin
Curt	12	oatmeal

The students want to put the cookies in bags. Each bag will contain all three kinds of cookies. All the bags will have the same of each kind of cookie.

a. Find all the ways each kind of cookies can be divided so there is the same number in each bag.

b. What is the greatest number they can put in each bag? How many bags of cookies does this make?

c. Other students want to bring in cookies. Name two other numbers of cookies that can be divided in the same way.

Show your work.

a. Bag1   Bag2   2 bags with 27 each
   12 P   12 P
   9 R   9 R
   6 O   6 O

   Bag1   Bag2   Bag3   3 bags with 18 each
   8 P   8 P   8 P
   6 R   6 R   6 R
   4 O   4 O   4 O

b. 2 bags would have the most of 27 cookies. They can't sell much this way.

c. If 2 bags a student could bring 30 cookies and have 15 in each bag.

Level 2 This work displays some misunderstanding of the task or concepts. Part *a* is incomplete. Part *b* is acceptable, but in part *c* only one additional number is given.

Model Student Papers for
Bake Sale

Name ______________________________

Bake Sale

Ariel, Brad, and Curt baked cookies for the school bake sale.

The students want to put the cookies in bags. Each bag will contain all three kinds of cookies. All the bags will have the same of each kind of cookie.

Baker	Number of Cookies	Kind of Cookie
Ariel	24	peanut
Brad	18	raisin
Curt	12	oatmeal

a. Find all the ways each kind of cookies can be divided so there is the same number in each bag.

b. What is the greatest number they can put in each bag? How many bags of cookies does this make? *The greatest number is 9 and it makes 6 bags.*

c. Other students want to bring in cookies. Name two other numbers of cookies that can be divided in the same way. *One of the numbers is 10 in each bag with is 60 cookies*

Show your work.

Level 1 This work displays minimal understanding of the task. The student gives one way of dividing the cookies.

Clowning Around

Name _______________

UNIT 5 • TASK B

Jolynn and her mother are making a clown costume for the class play. They found some pieces of fabric at a yard sale. Help them choose fabric pieces for the costume

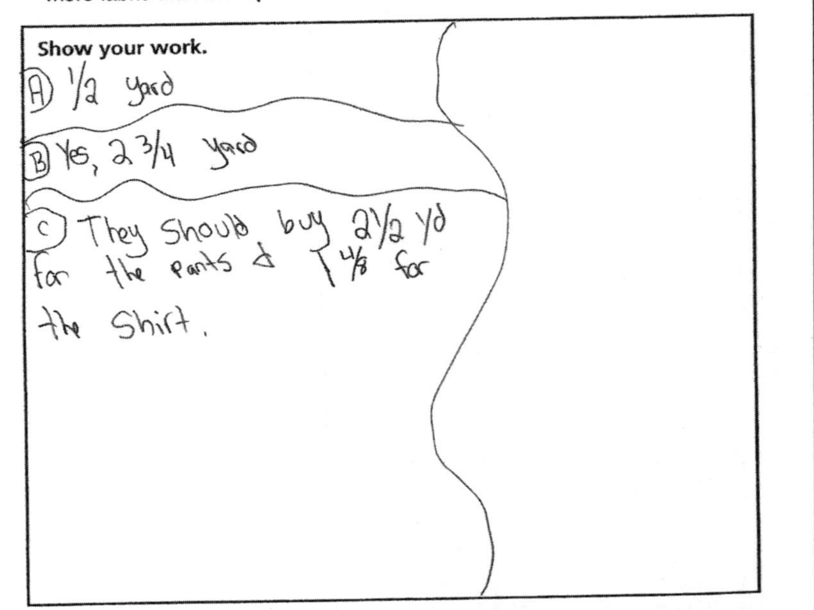

a. They can use the shortest piece of fabric for the hat. Which piece should they choose for the hat?

b. They need $2\frac{5}{8}$ yards for the top of the costume. Is there a piece large enough for the top? If so, which is it?

c. Decide what other pieces of fabric they should buy for the sleeves and the pants of the costume. The pants will take more fabric than the top or the sleeves.

Show your work.

a. $\frac{1}{2}$ yard is the shortest.

b. Yes, it is the $2\frac{3}{4}$ piece. $2\frac{3}{4} = 2\frac{6}{8}$ so you will have $\frac{1}{8}$ yard left over.

c. For the sleeves I would pick the $1\frac{1}{2}$ yards piece. This seems big enough for sleeves. For the pants I need more than $2\frac{5}{8}$ yards. I can get $4\frac{3}{4}$ yards if I buy the $2\frac{1}{2} + 2\frac{2}{8} = 4\frac{6}{8} = 4\frac{3}{4}$ yards, I picked the two white pieces so the legs will match.

Level 3 This student shows a good understanding of the requirements of the task. The work is logical and accurate. Explanations are given with the choices.

Clowning Around

Name _______________

UNIT 5 • TASK B

Jolynn and her mother are making a clown costume for the class play. They found some pieces of fabric at a yard sale. Help them choose fabric pieces for the costume

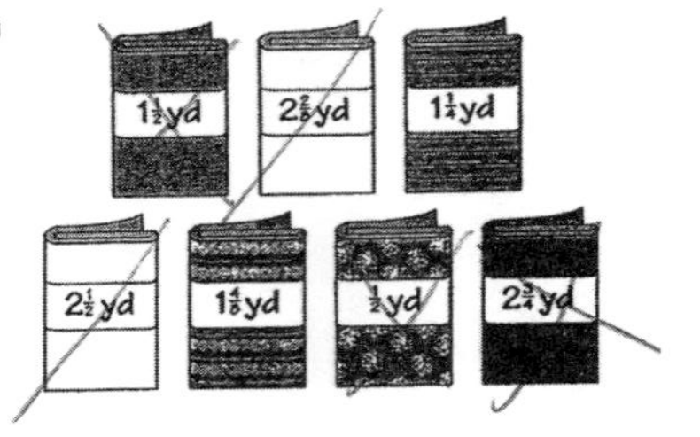

a. They can use the shortest piece of fabric for the hat. Which piece should they choose for the hat?

b. They need $2\frac{5}{8}$ yards for the top of the costume. Is there a piece large enough for the top? If so, which is it?

c. Decide what other pieces of fabric they should buy for the sleeves and the pants of the costume. The pants will take more fabric than the top or the sleeves.

Show your work.

A) ½ yard

B) Yes, 2 3/4 yard

C) They should buy 2½ yd for the pants & 1⅘ for the shirt.

Level 2 Student displays good understanding of the task. Parts *a* and *b* are correct; part c is not correct.

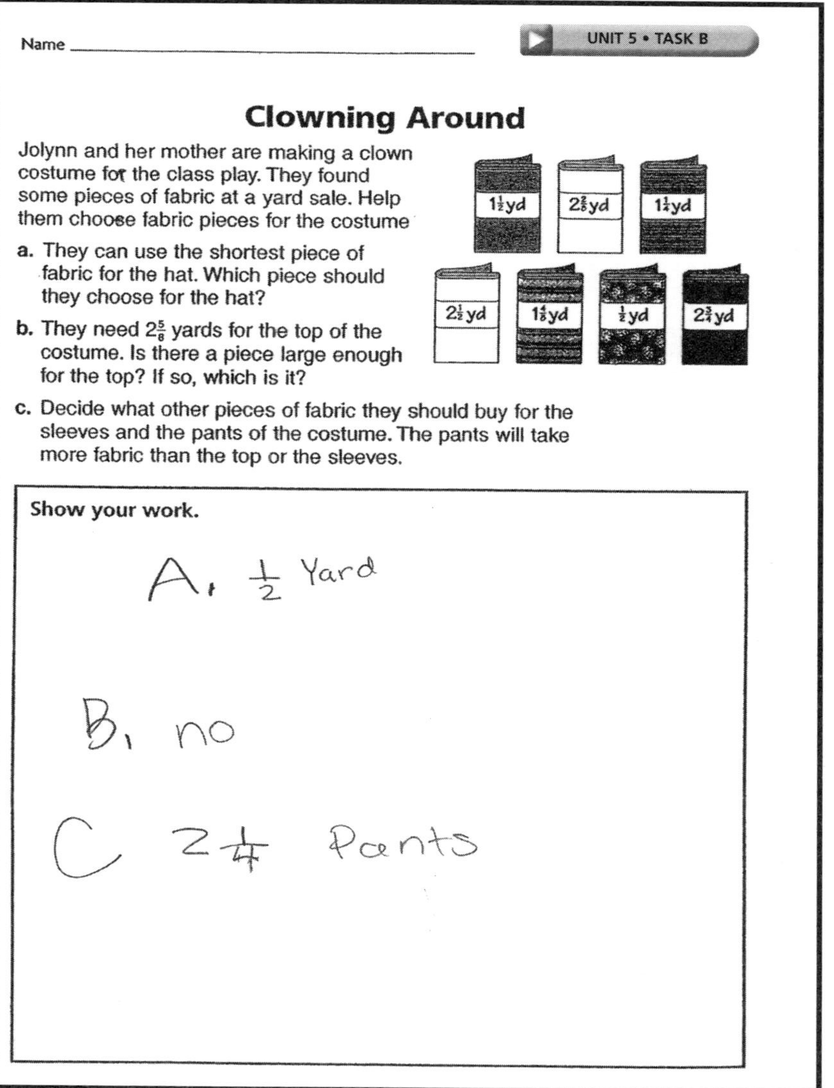

Name ______________________

Clowning Around

Jolynn and her mother are making a clown costume for the class play. They found some pieces of fabric at a yard sale. Help them choose fabric pieces for the costume

a. They can use the shortest piece of fabric for the hat. Which piece should they choose for the hat?

b. They need $2\frac{5}{8}$ yards for the top of the costume. Is there a piece large enough for the top? If so, which is it?

c. Decide what other pieces of fabric they should buy for the sleeves and the pants of the costume. The pants will take more fabric than the top or the sleeves.

Show your work.

A. ½ Yard

B. no

C 2¼ Pants

Level 1 The response shows limited understanding. Part *a* is correct; parts *b* and *c* are inadequate and inaccurate.

Pizza Time

TASK A

Purpose
To assess student understanding of modeling and addition of fractions and mixed numbers

Grouping
Individuals

Time
5–10 minutes

Preparation Hints
Discuss the number of slices for the pizza. Mention that some questions may have more than one answer.

Introduce the Task
Students solve a problem about sharing a pizza. The shares are not equal, but must fit conditions of the problem.

On the Job

TASK B

Purpose
To assess student understanding of addition and multiplication of fractions and mixed numbers

Grouping
Individuals

Time
10–15 minutes

Introduce the Task
Students decide which part-time jobs fit students' schedules. They must consider different conditions to find a match.

TASK A

Pizza Time

Performance Indicators	Observations and Rubric Score
________ Draws a diagram to show parts of a whole. ________ Writes fractions to describe parts of a whole. ________ Solves a problem involving mixed numbers to fit a set of conditions. ________ Shows work and explains how the answers were determined.	3 2 1 0

TASK B

On the Job

Performance Indicators	Observations and Rubric Score
________ Selects a job to fit a schedule. ________ Selects a job to fit day available to work and pay desired. ________ Changes job descriptions to allow choices. ________ Shows work and explains how the answers were determined.	3 2 1 0

Total Score _________ /6

Pizza Time

Dan bought a pizza to share with Kenny. Each boy ate a different number of pieces. Dan ate the most. When the boys were finished there was one piece left for Dan to give to his grandmother.

a. Draw a diagram of the pizza. Write fractions on the diagram to show the amounts Dan and Kenny ate.

b. If the boys ate the amounts you suggested, what fraction of the pizza was left for Dan's grandmother?

c. Dan baked four strawberry tarts to share with his grandmother and Kenny. Use mixed numbers to suggest a way they could have shared the tarts if Dan ate the most and his grandmother ate the least.

Show your work.

On the Job

A local grocery store has these part-time jobs available. Jon and Kate each want to apply for one of these jobs.

- HELP WANTED -	- HELP WANTED -	- HELP WANTED -
Job A	*Job B*	*Job C*
Stock shelves after school for $2\frac{1}{2}$ hours per day, 4 days a week. Pays $6 per hour.	Help at check-out counters for $3\frac{1}{2}$ hours every Friday and $5\frac{1}{2}$ hours every Saturday. Pays $6 per hour.	Unload 3 delivery trucks every Saturday. Each truck requires about $2\frac{1}{2}$ hours to unload. Pays $20 per truck.

a. Jon wants to work 10 hours per week. Which job should he apply for?

b. Kate wants to work only on Saturdays to earn $100 as quickly as possible. Which job should she apply for?

c. Change the job descriptions so that Jon and Kate can choose between two part-time jobs.

Show your work.

Pizza Time

Name ________________

Dan bought a pizza to share with Kenny. Each boy ate a different number of pieces. Dan ate the most. When the boys were finished there was one piece left for Dan to give to his grandmother.

a. Draw a diagram of the pizza. Write fractions on the diagram to show the amounts Dan and Kenny ate.

b. If the boys ate the amounts you suggested, what fraction of the pizza was left for Dan's grandmother?

c. Dan baked four strawberry tarts to share with his grandmother and Kenny. Use mixed numbers to suggest a way they could have shared the tarts if Dan ate the most and his grandmother ate the least.

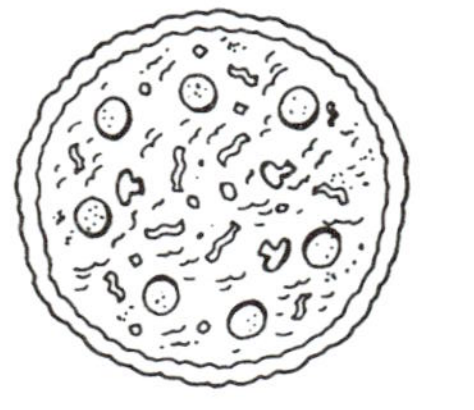

Show your work.

a. Students' diagrams should show a pizza divided into equal parts with 1 piece for grandmother and more pieces for Dan than Kenny.
Sample answer:
The diagram could be divided into 10 pieces with 5 pieces for Dan, 4 pieces for Kenny, and 1 piece for grandmother.
fraction for Dan: $\frac{5}{10}$, or $\frac{1}{2}$
fraction for Kenny: $\frac{4}{10}$, or $\frac{2}{5}$
b. fraction for grandmother: $\frac{1}{10}$

c. Sample answer for sharing tarts:

Dan	2	$1\frac{3}{4}$	2
Kenny	$1\frac{1}{2}$	$1\frac{1}{4}$	$1\frac{1}{4}$
Grandmother	$\frac{1}{2}$	1	$\frac{3}{4}$

PA48 Performance Assessment

On the Job

Name ________________

A local grocery store has these part-time jobs available. Jon and Kate each want to apply for one of these jobs.

- HELP WANTED -
Job A
Stock shelves after school for $2\frac{1}{2}$ hours per day, 4 days a week. Pays $6 per hour.

- HELP WANTED -
Job B
Help at check-out counters for $3\frac{1}{2}$ hours every Friday and $5\frac{1}{2}$ hours every Saturday. Pays $6 per hour.

- HELP WANTED -
Job C
Unload 3 delivery trucks every Saturday. Each truck requires about $2\frac{1}{2}$ hours to unload. Pays $20 per truck.

a. Jon wants to work 10 hours per week. Which job should he apply for? To work 10 hours per week, Jon should apply for Job A (4 days at $2\frac{1}{2}$ hours per day is 10 hours).

b. Kate wants to work only on Saturdays to earn $100 as quickly as possible. Which job should she apply for? Kate should apply for Job C to work only Saturdays and to earn $100 as quickly as possible.

c. Change the job descriptions so that Jon and Kate can choose between two part-time jobs.

Show your work.

Sample answer:
To change job descriptions so Jon can choose between two jobs: for Job B increase Friday hours to $4\frac{1}{2}$; or for Job C, increase number of trucks to 4.
For Kate to choose between two jobs: change days for A or B to Saturdays only.
So both Jon and Kate can choose between 2 part-time jobs, change job B description to $4\frac{1}{2}$ hours on Friday and $5\frac{1}{2}$ hours on Saturday or $8\frac{1}{2}$ hours on Saturday.

Performance Assessment PA49

Model Student Papers for
Pizza Time

Pizza Time

Dan bought a pizza to share with Kenny. Each boy ate a different number of pieces. Dan ate the most. When the boys were finished there was one piece left for Dan to give to his grandmother.

a. Draw a diagram of the pizza. Write fractions on the diagram to show the amounts Dan and Kenny ate.

b. If the boys ate the amounts you suggested, what fraction of the pizza was left for Dan's grandmother?

c. Dan baked four strawberry tarts to share with his grandmother and Kenny. Use mixed numbers to suggest a way they could have shared the tarts if Dan ate the most and his grandmother ate the least.

Show your work.

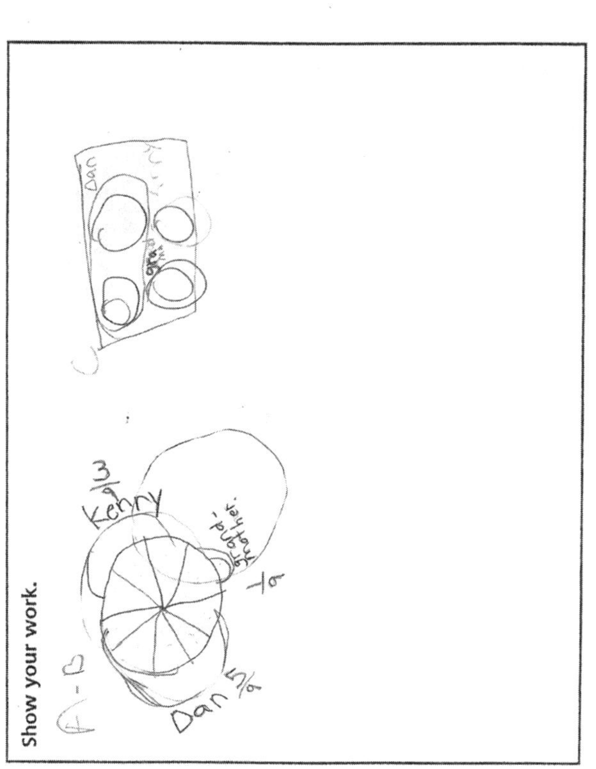

Level 2 Evidence of understanding is limited. Diagrams and fractions are correct, but there is little explanation especially in part c.

Pizza Time

Dan bought a pizza to share with Kenny. Each boy ate a different number of pieces. Dan ate the most. When the boys were finished there was one piece left for Dan to give to his grandmother.

a. Draw a diagram of the pizza. Write fractions on the diagram to show the amounts Dan and Kenny ate.

b. If the boys ate the amounts you suggested, what fraction of the pizza was left for Dan's grandmother?

c. Dan baked four strawberry tarts to share with his grandmother and Kenny. Use mixed numbers to suggest a way they could have shared the tarts if Dan ate the most and his grandmother ate the least.

Show your work.

Level 3 This student shows good understanding of fractions. Answers are complete and accurate. All parts are complete.

Name ___________________________

UNIT 6

Pizza Time

Dan bought a pizza to share with Kenny. Each boy ate a different number of pieces. Dan ate the most. When the boys were finished there was one piece left for Dan to give to his grandmother.

a. Draw a diagram of the pizza. Write fractions on the diagram to show the amounts Dan and Kenny ate.

b. If the boys ate the amounts you suggested, what fraction of the pizza was left for Dan's grandmother?

c. Dan baked four strawberry tarts to share with his grandmother and Kenny. Use mixed numbers to suggest a way they could have shared the tarts if Dan ate the most and his grandmother ate the least.

Show your work.

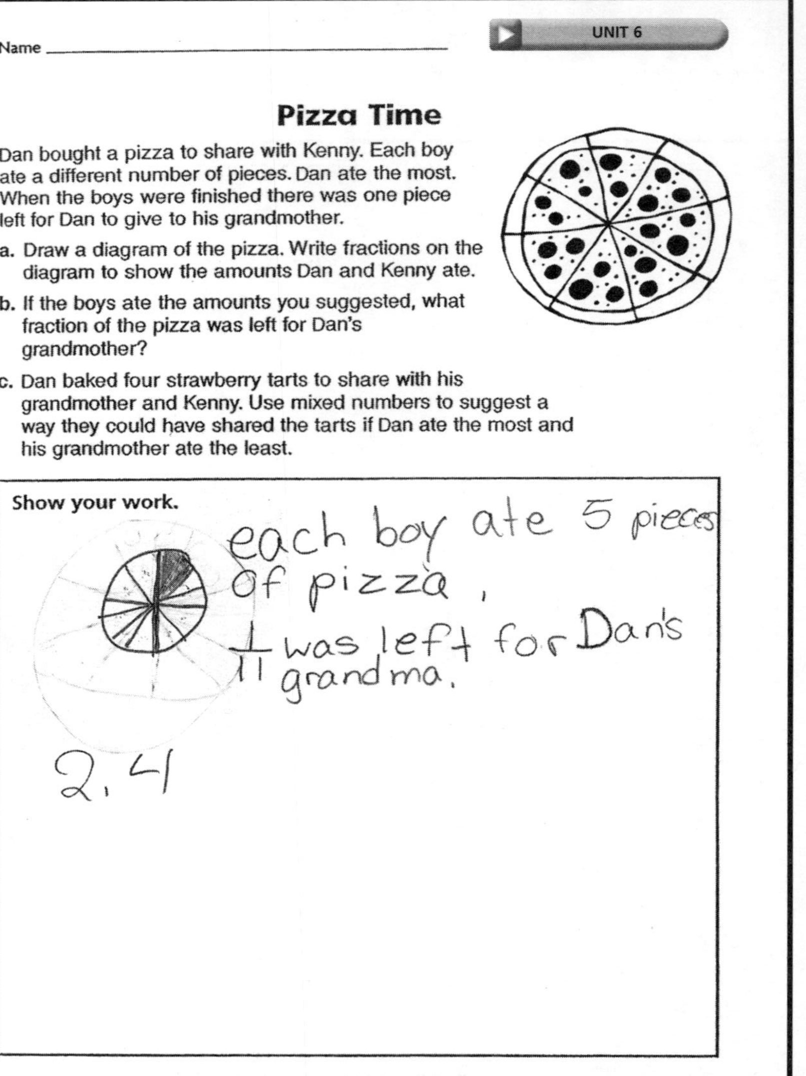

Level 1 Work shown is incomplete. Part *a* shows no use of fractions. Part *b* is correct; part *c* is unexplained and mixed numbers are not used.

Level 3 (left paper)

Name ______________________

UNIT 6 • TASK B

On the Job

A local grocery store has these part-time jobs available. Jon and Kate each want to apply for one of these jobs.

- HELP WANTED -
Job A
Stock shelves after school for $2\frac{1}{2}$ hours per day, 4 days a week. Pays $6 per hour.

- HELP WANTED -
Job B
Help at check-out counters for $3\frac{1}{2}$ hours every Friday and $5\frac{1}{2}$ hours every Saturday. Pays $6 per hour.

- HELP WANTED -
Job C
Unload 3 delivery trucks every Saturday: Each truck requires about $2\frac{1}{2}$ hours to unload. Pays $20 per truck.

a. Jon wants to work 10 hours per week. Which job should he apply for? A

b. Kate wants to work only on Saturdays to earn $100 as quickly as possible. Which job should she apply for? C

c. Change the job descriptions so that Jon and Kate can choose between two part-time jobs.

Show your work.
a. Job A is 10 hours $4 \times 2\frac{1}{2} = 10$
B is 9 hours $5 + 3 + \frac{1}{2} + \frac{1}{2} = 9$
C is $7\frac{1}{2}$ hours $2 \times 3 = 6 + \frac{1}{2} + \frac{1}{2} + \frac{1}{2}$
Jon needs Job A.

b. Job B – Saturday for $5\frac{1}{2}$ hours for $6 = 30 + 3 = \$33$
Job C – Saturday only – 3 trucks at $20 = \$60$
Kate has to get Job C.

c. Job C – HELP!
Unload 4 trucks on Saturday. Each truck $2\frac{1}{2}$ hours. $20 each.

Job B –
Help check out for $3\frac{1}{2}$ hours Friday or for $7\frac{1}{2}$ hours Saturday. Earn $8.50 an hour.

Jon can do Job C now for 10 hours.

Now Job B is Saturday for Kate.

Level 3   This student demonstrates understanding of the task. Computations are correct and work is logical. Parts *a* and *b* are accurate and complete. Part *c* could be more complete.

Level 2 (right paper)

Name ______________________

UNIT 6 • TASK B

On the Job

A local grocery store has these part-time jobs available. Jon and Kate each want to apply for one of these jobs.

- HELP WANTED -
Job A
Stock shelves after school for $2\frac{1}{2}$ hours per day, 4 days a week. Pays $6 per hour.

- HELP WANTED -
Job B
Help at check-out counters for $3\frac{1}{2}$ hours every Friday and $5\frac{1}{2}$ hours every Saturday. Pays $6 per hour.

- HELP WANTED -
Job C
Unload 3 delivery trucks every Saturday: Each truck requires about $2\frac{1}{2}$ hours to unload. Pays $20 per truck.

a. Jon wants to work 10 hours per week. Which job should he apply for? Job A

b. Kate wants to work only on Saturdays to earn $100 as quickly as possible. Which job should she apply for? Job C

c. Change the job descriptions so that Jon and Kate can choose between two part-time jobs.

Show your work.
a) A - $2\frac{1}{4} \times 4 = 8 + 2 = 10$ hours   B - $5\frac{1}{2} + 3\frac{1}{2} = 9$ hours
C - $2\frac{1}{2} + 2\frac{1}{2} + 2\frac{1}{2} = 7\frac{1}{2}$

b.) C – Saturday only – $60 for a Saturday – This is more money than the other jobs

c.) change B & C

Level 2   The work shown is correct but incomplete. Parts *a* and *b* are accurate; part *c* is incomplete.

On the Job

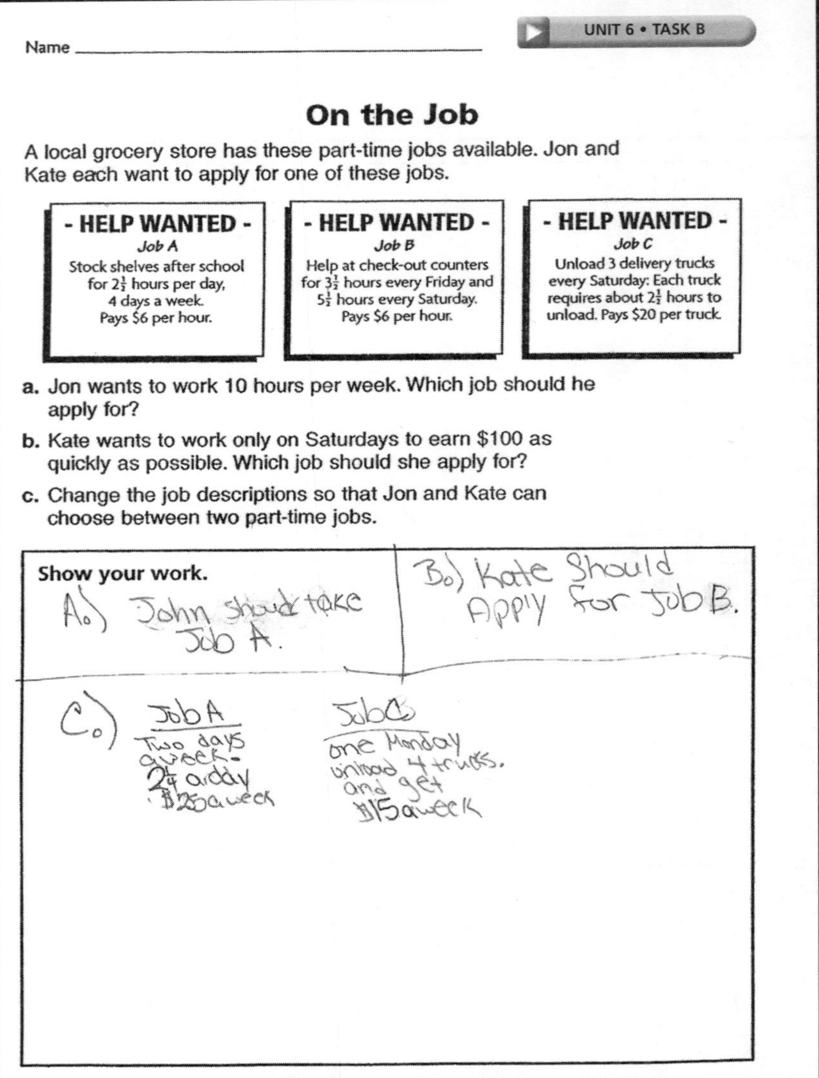

Level 1   This student demonstrates minimal understanding of the task and concept. Only part *a* is correct. Parts *b* and *c* are incorrect.

TASK A

The Winning Design

Purpose
To assess student understanding of similar and congruent figures and rotational symmetry

Materials
ruler, compass, protractor

Grouping
Individuals

Time
10–15 minutes

Preparation Hints
Review what is meant by rotational symmetry.

Introduce the Task
Students design a contest poster for an art fair. Students must identify congruent figures and rotational symmetry in their designs.

TASK B

Naming Ordered Pairs

Purpose
To assess student understanding of using ordered pairs to locate points on a grid

Materials
rulers

Grouping
Individuals

Time
15–20 minutes

Preparation Hints
Establish the lower left corner as the origin (0,0) for a first quadrant coordinate system. To keep corners of furniture on lattice points, declare that the furniture must be parallel to the side of the grid. Remind students of the order in ordered pairs.

Introduce the Task
Students will arrange furniture on a grid. They must pinpoint the locations of the corners of the furniture pieces using ordered pairs.

TASK A # The Winning Design

Performance Indicators	Observations and Rubric Score
_________ Designs a poster using circles and similar and congruent figures. _________ Identifies rotational symmetry in the figures. _________ Identifies congruent figures in the design. _________ Shows work and explains how the answers were determined.	3 2 1 0

TASK B # Naming Ordered Pairs

Performance Indicators	Observations and Rubric Score
_________ Identifies ordered pairs to show the location of furniture. _________ Moves and draws congruent figures on a grid. _________ Identifies ordered pairs to show the new location of furniture. _________ Shows work and explains how the answers were determined.	3 2 1 0

Total Score _________ /6

The Winning Design

The math club is having a poster contest. The winning design will be used on posters to advertise the Spring Art Fair. These are the rules for the design:

1. The design should fit on a 24-inch by 24-inch poster.

2. Only a ruler, compass, protractor, and pencil may be used to create the design.

3. A circle must be the first figure drawn.

4. There must be some rotational symmetry in the design.

5. There must be similar and congruent figures in the design.

a. Create a design for the poster. Then write descriptions of shapes and sizes in the design so that someone else could make the same design.

b. The rules ask you to show rotational symmetry in your design. What did you draw?

c. Identify the figures in your design that are congruent.

Show your work.

Naming Ordered Pairs

The map shows where some of the furniture in
Mrs. Henderson's classroom is located. She
has asked her students to draw a new
arrangement for the furniture.

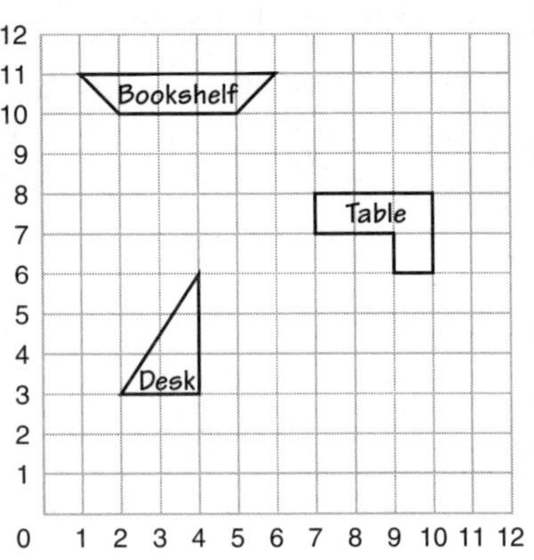

a. Write the ordered pairs to
show the locations for the
corners of the bookshelf, the
desk, and the table as they
appear on the map.

____

____

b. Draw the new location for each piece of furniture on the grid.

____

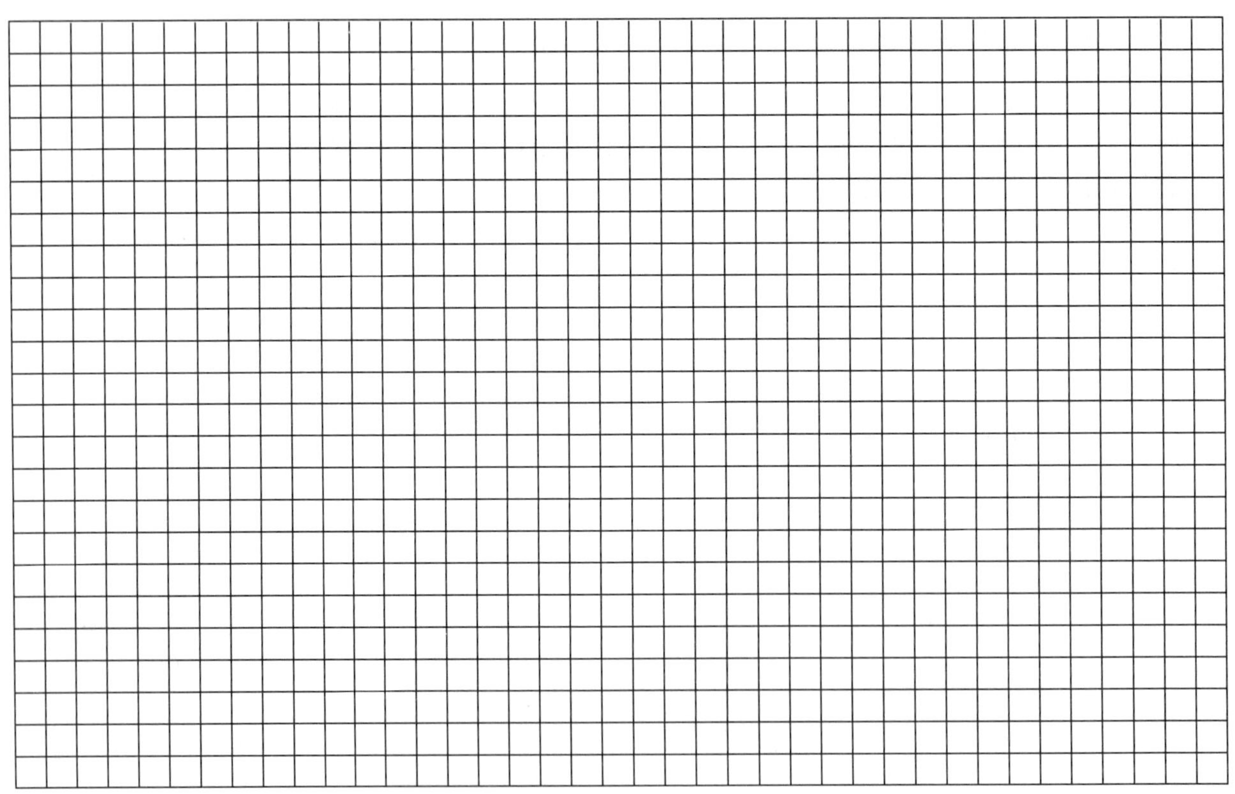

c. Name the ordered pairs to show the new locations of the
bookshelf, the desk, and the table.

____

PA58   Performance Assessment

The Winning Design

UNIT 7 • TASK A

Name ___________________________

The math club is having a poster contest. The winning design will be used on posters to advertise the Spring Art Fair. These are the rules for the design:

1. The design should fit on a 24-inch by 24-inch poster

2. Only a ruler, compass, protractor, and pencil may be used to create the design.

3. A circle must be the first figure drawn.

4. There must be some rotational symmetry in the design.

5. There must be similar and congruent figures in the design.

a. Create a design for the poster. Then write descriptions of Students' designs will shapes and sizes in the design so that someone else could vary. The student's make the same design. design must have: a circle, rotational symmetry, similar figures, and congruent figures. Students should write descriptions of shapes in the design.

b. The rules ask you to show rotational symmetry in your design. What did you draw? Students should also identify the rotational symmetry and congruent figures in their design.

c. Identify the figures in your design that are congruent. Check student's posters.

Show your work.

Performance Assessment **PA**57

Naming Ordered Pairs

UNIT 7 • TASK B

Name ___________________________

The map shows where some of the furniture in Mrs. Henderson's classroom is located. She has asked her students to draw a new arrangement for the furniture.

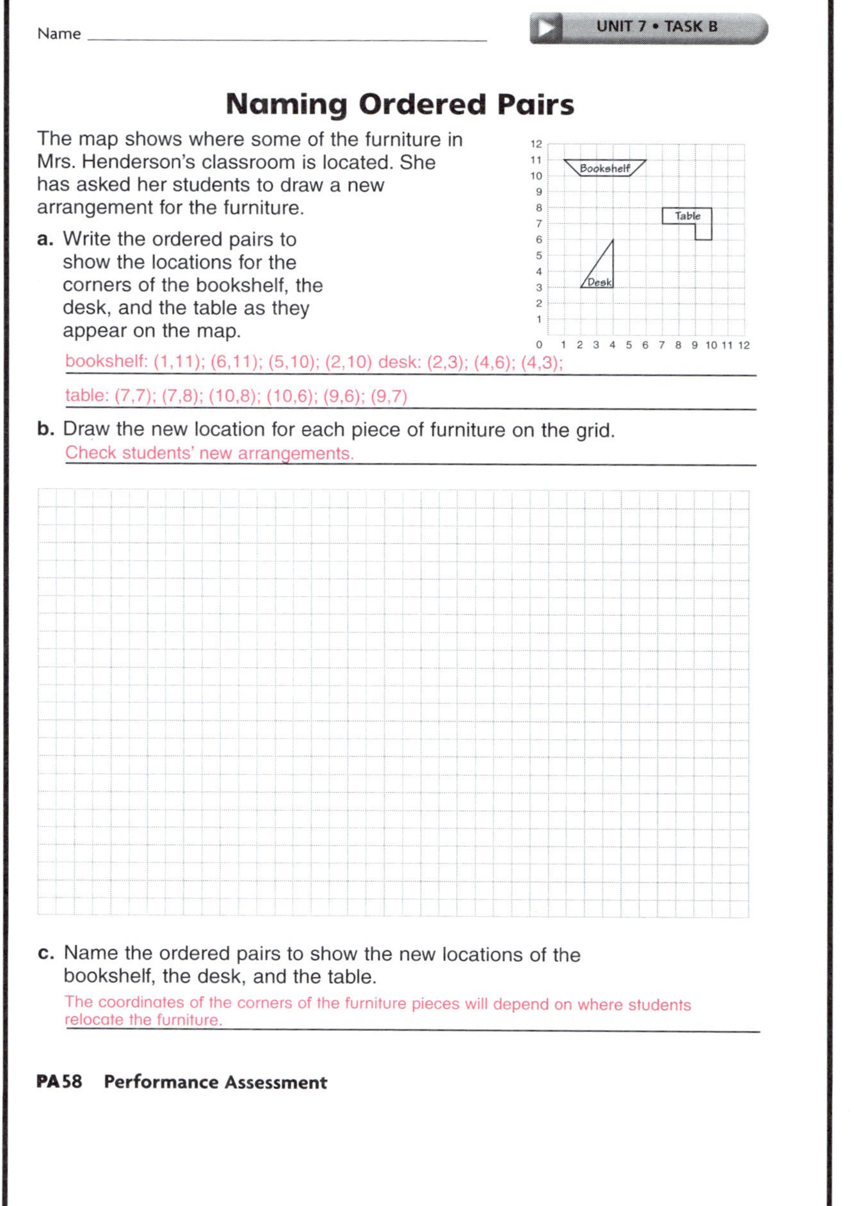

a. Write the ordered pairs to show the locations for the corners of the bookshelf, the desk, and the table as they appear on the map.

bookshelf: (1,11); (6,11); (5,10); (2,10) desk: (2,3); (4,6); (4,3);

table: (7,7); (7,8); (10,8); (10,6); (9,6); (9,7)

b. Draw the new location for each piece of furniture on the grid.
Check students' new arrangements.

c. Name the ordered pairs to show the new locations of the bookshelf, the desk, and the table.

The coordinates of the corners of the furniture pieces will depend on where students relocate the furniture.

PA58 Performance Assessment

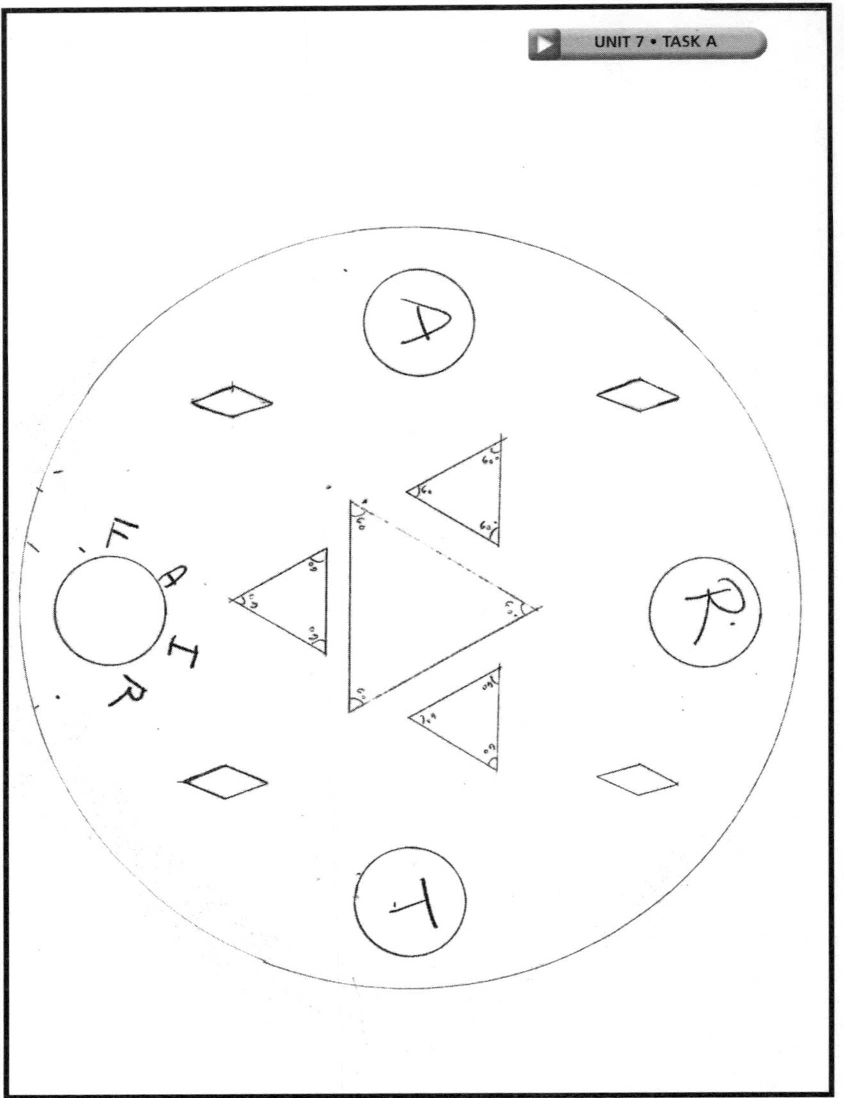

(Additional page of student sample paper)

Name ___________________

The Winning Design

The math club is having a poster contest. The winning design will be used on posters to advertise the Spring Art Fair. These are the rules for the design:

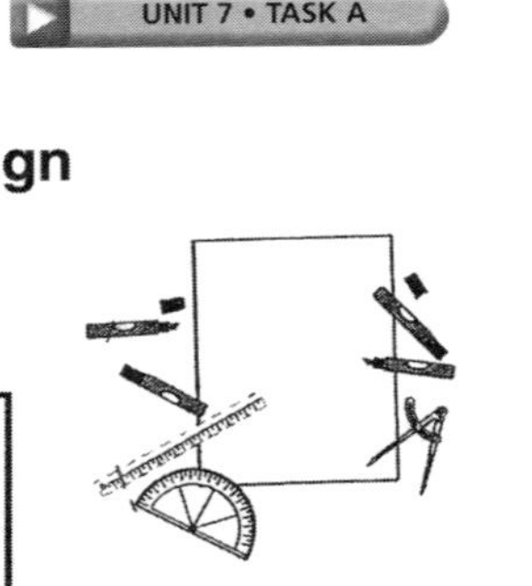

1. The design should fit on a 24-inch by 24-inch poster.

2. Only a ruler, compass, protractor, and pencil may be used to create the design.

3. A circle must be the first figure drawn.

4. There must be some rotational symmetry in the design.

5. There must be similar and congruent figures in the design.

a. Create a design for the poster. Then write descriptions of shapes and sizes in the design so that someone else could make the same design.

b. The rules ask you to show identify rotational symmetry in your design. What did you draw?

c. Identify the figures in your design that are congruent. The three small triangles are. The four small circles are. The rhombuses are.

Show your work.

a. The design is a big circle. I put four evenly spaced circles around the rim of the big circle. In the center of the big circle is an equilateral triangle with each side of 5 cm. I was careful to make each angle 60°. I put three smaller equilateral triangles against each side of the big triangle. Each side is ½ the side of the big triangle. Last I put four rhombuses. These were the hardest to get the angles right.

b. Each of my shapes has rotational symmetry — equilateral triangle, rhombus, circle. They look the same in two positions when you rotate them.

Level 3   The student shows a good understanding of the task. Figures are accurate and explanations show appropriate understanding.

Left paper (Level 2)

Name ________________________

The Winning Design

The math club is having a poster contest. The winning design will be used on posters to advertise the Spring Art Fair. These are the rules for the design:

1. The design should fit on a 24-inch by 24-inch poster.
2. Only a ruler, compass, protractor, and pencil may be used to create the design.
3. A circle must be the first figure drawn.
4. There must be some rotational symmetry in the design.
5. There must be similar and congruent figures in the design.

a. Create a design for the poster. Then write descriptions of shapes and sizes in the design so that someone else could make the same design.

b. The rules ask you to show identify rotational symmetry in your design. What did you draw?

c. Identify the figures in your design that are congruent.

Show your work. (A) There are two triangles, 6 circles and 4 rectangles. There
(B) All of my shapes have are 4 congruent circles and two congrent squares,
rotational symmetry.
(C) There are 4 congrent circles and two congruent squares.

Level 2 This work lacks the specificity and accuracy of a level 3 paper. Some confusion exists as to congruent circles and rotational symmetry. Some figures lack precision.

Right paper (Level 1)

Name ________________________

The Winning Design

The math club is having a poster contest. The winning design will be used on posters to advertise the Spring Art Fair. These are the rules for the design:

1. The design should fit on a 24-inch by 24-inch poster.
2. Only a ruler, compass, protractor, and pencil may be used to create the design.
3. A circle must be the first figure drawn.
4. There must be some rotational symmetry in the design.
5. There must be similar and congruent figures in the design.

a. Create a design for the poster. Then write descriptions of shapes and sizes in the design so that someone else could make the same design. Squares and a circle

b. The rules ask you to show identify rotational symmetry in your design. What did you draw? circle

c. Identify the figures in your design that are congruent. squares

Show your work.

Level 1 A minimal response is provided. There is a major error in that the squares are not congruent.

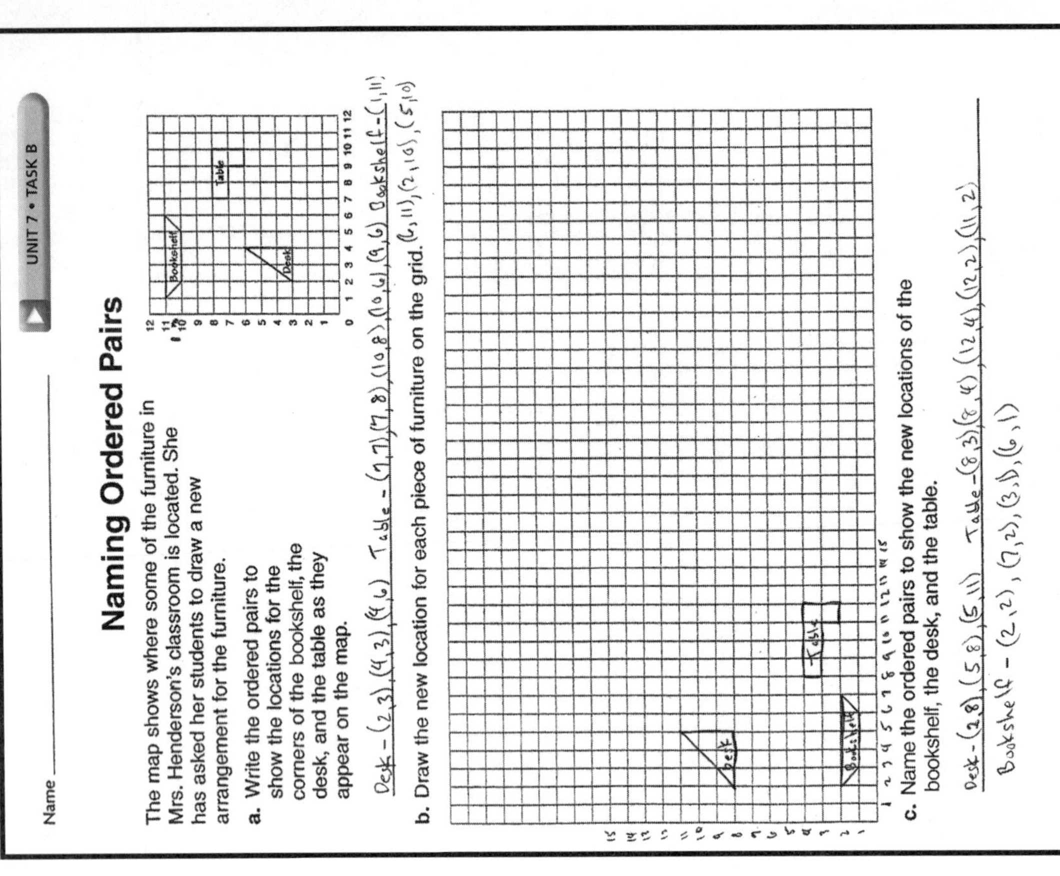

Level 2 This work is mostly correct. It is accurate except for one missing ordered pair for the table. Furniture has been moved, but desk and table are now larger than originals.

Level 3 This student shows understanding of the task and an understanding of locating points on a grid. Answers are complete and accurate.

Naming Ordered Pairs

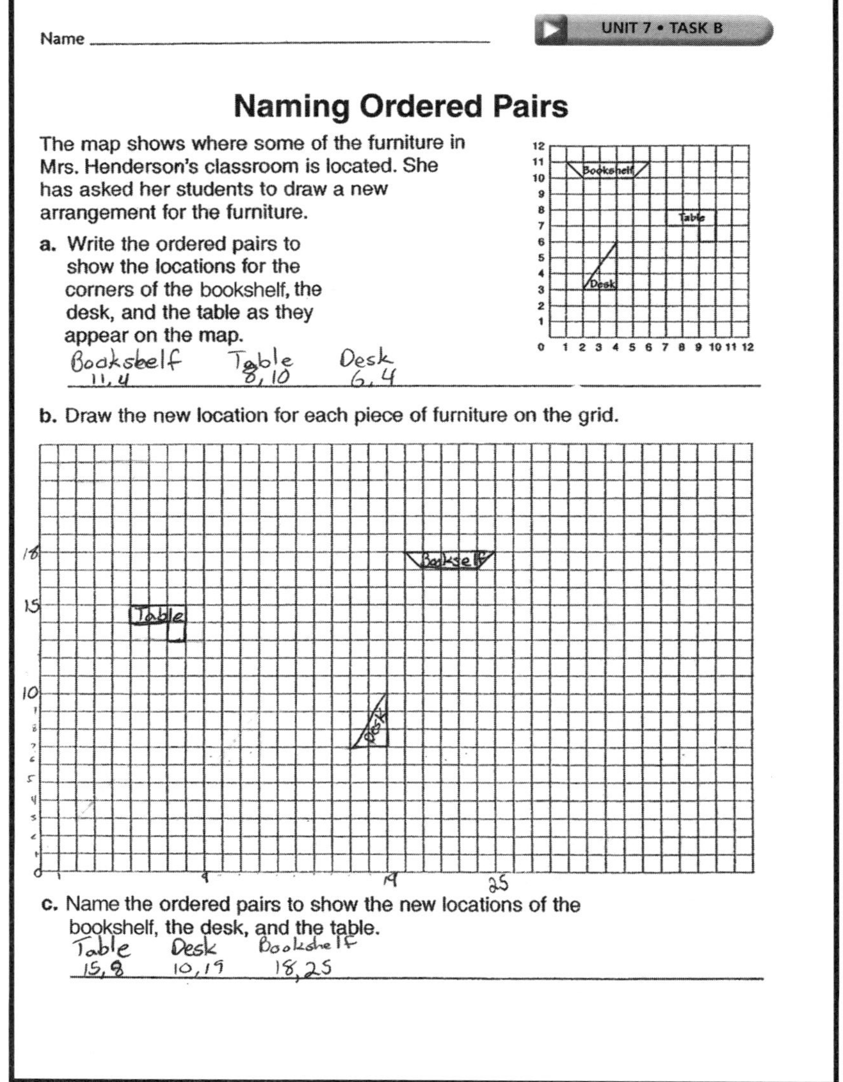

Name ___________

Naming Ordered Pairs

The map shows where some of the furniture in Mrs. Henderson's classroom is located. She has asked her students to draw a new arrangement for the furniture.

a. Write the ordered pairs to show the locations for the corners of the bookshelf, the desk, and the table as they appear on the map.

Bookshelf   Table   Desk
11,4   8,10   6,4

b. Draw the new location for each piece of furniture on the grid.

c. Name the ordered pairs to show the new locations of the bookshelf, the desk, and the table.

Table   Desk   Bookshelf
15,8   10,19   18,25

Level 1   This work is incomplete. New locations are drawn for the furniture and the furniture is the right size; however not all ordered pairs are given.

TASK A

Woodworking

Purpose
To assess student understanding of areas of triangles and rectangles

Materials
centimeter ruler and grid paper

Grouping
Individuals

Time
15–20 minutes

Preparation Hints
Using grid paper with 1-cm squares will make measuring dimensions and finding areas simpler.

Introduce the Task
Students design the top of a jewelry box using triangles and quadrilaterals on grid paper. They must measure dimensions of figures and find the areas of figures.

TASK B

Storage Cabinet

Purpose
To assess student understanding of perimeter, area, and volume

Grouping
Individuals or partners

Time
15–20 minutes

Preparation Hints
Review the differences in perimeter, area, and volume and their units of measurement.

Introduce the Task
Students plan the dimensions of a storage cabinet. They must calculate perimeter, area, and volume to decide how much material is needed for paint and trim.

TASK A

Woodworking

Performance Indicators	Observations and Rubric Score
_________ Draws a figure less than 15 cm by 15 cm with triangles, rectangles, and parallelograms. _________ Finds the areas of the largest triangle and rectangle. _________ Calculates the total price given the price per unit of area. _________ Shows work and explains how the answers were determined.	3 2 1 0

TASK B

Storage Cabinet

Performance Indicators	Observations and Rubric Score
_________ Chooses dimensions to produce a given volume for a cabinet. _________ Finds the surface area of the cabinet to decide if one can of paint is enough. _________ Finds the perimeter of the top of the cabinet to decide how much trim is needed. _________ Shows work and explains how the answers were determined.	3 2 1 0

Total Score _________/6

Woodworking

Materials: centimeter ruler

Mr. Leon makes wooden jewelry boxes. On the tops of the boxes, he uses different kinds of wood to create geometric designs with triangles and quadrilaterals.

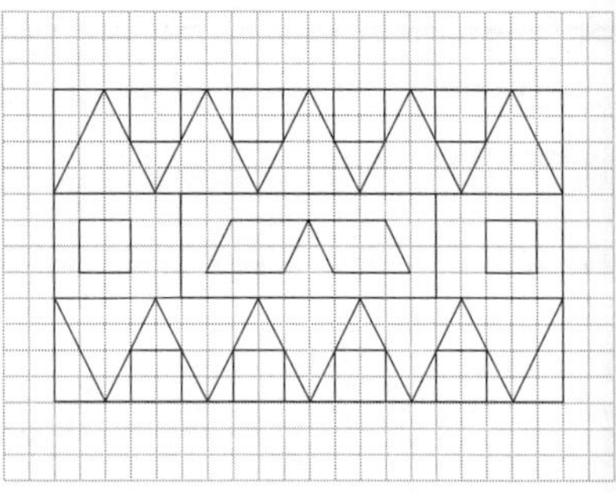

a. Draw a figure that is less than 15 centimeters by 15 centimeters. Use the figure to design a box top that includes triangles, rectangles, and parallelograms.

b. Use a centimeter ruler to measure the dimensions of the largest triangle and rectangle in your design. Write the dimensions on the diagram. Then find the areas of those figures.

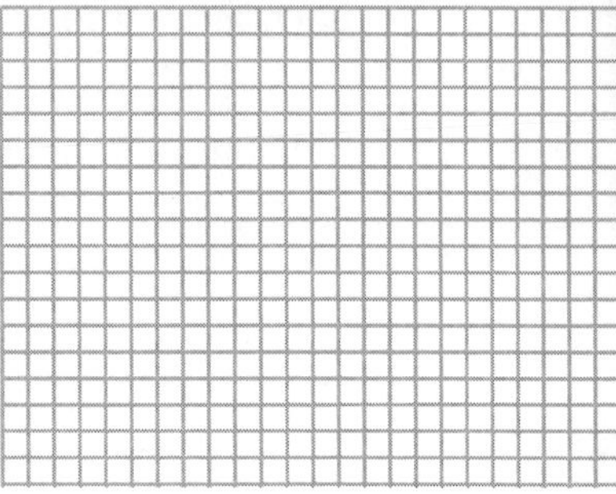

c. Mr. Leon charges $0.25 per square centimeter to make his box tops. How much would he charge for the box you designed?

Show your work.

Storage Cabinet

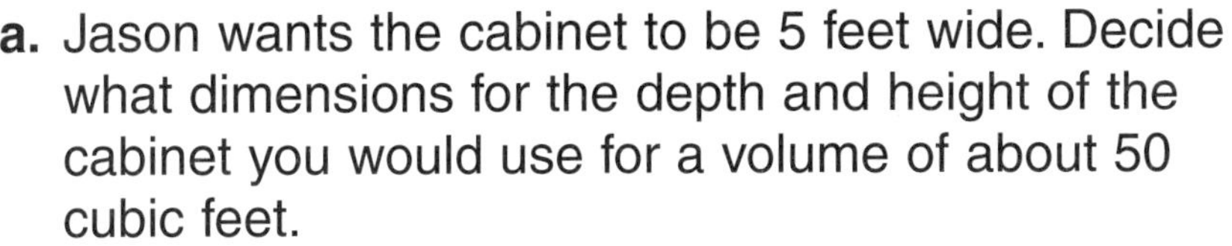

Jason and his mother are designing a cabinet to store his sports equipment. They want the cabinet to have about 50 cubic feet of space.

a. Jason wants the cabinet to be 5 feet wide. Decide what dimensions for the depth and height of the cabinet you would use for a volume of about 50 cubic feet.

b. Jason wants to paint the cabinet. He reads on the paint can that the paint will cover 100 square feet. Does he have enough paint to cover the back, sides, and the top of the cabinet?

c. Suppose Jason decides to put a wooden ledge around the top of the cabinet. How many feet of trim will he need?

Show your work.

Name _______________________

Woodworking

Materials: centimeter ruler

Mr. Leon makes wooden jewelry boxes. On the tops of the boxes, he uses different kinds of wood to create geometric designs with triangles and quadrilaterals.

a. Draw a figure that is less than 15 centimeters by 15 centimeters. Use the figure to design a box top that includes triangles, rectangles, and parallelograms.

b. Use a centimeter ruler to measure the dimensions of the largest triangle and rectangle in your design. Write the dimensions on the diagram. Then find the areas of those figures.

c. Mr. Leon charges $0.25 per square centimeter to make his box tops. How much would he charge for the box you designed?

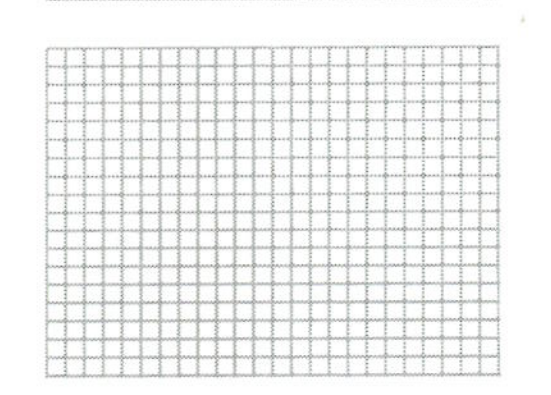

Show your work.

Students' designs for box tops must be less than 15 cm by 15 cm and be composed of triangles, rectangles, and parallelograms.

Each student will find the areas of the largest rectangle and largest triangle of his or her design.

Students need to find the area of their box tops in square centimeters. To find the cost to have their designs made into a box top, multiply area by $0.25 or divide area by 4 to get the charge in dollars.

Name _______________________

Storage Cabinet

Jason and his mother are designing a cabinet to store his sports equipment. They want the cabinet to have about 50 cubic feet of space.

a. Jason wants the cabinet to be 5 feet wide. Decide what dimensions for the depth and height of the cabinet you would use for a volume of about 50 cubic feet.

b. Jason wants to paint the cabinet. He reads on the paint can that the paint will cover 100 square feet. Does he have enough paint to cover the back, sides, and the top of the cabinet?

c. Suppose Jason decides to put a wooden ledge around the top of the cabinet. How many feet of trim will he need?

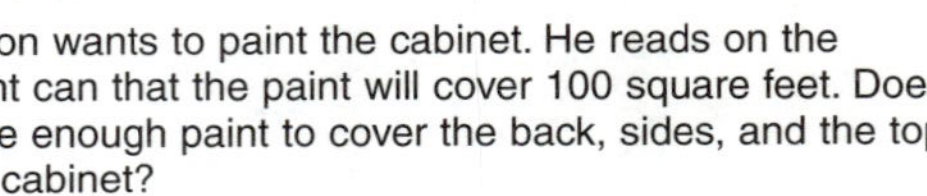

Show your work.

Students' choices for the depth and height of the cabinet can vary, but the product of their depth and height must be about 10.

Samples: 5 by 5 by 2
          5 by 4 by 2.5

The amount of area to be painted depends on the student's choice for height and depth. To compute area to be painted:

		$d = 2, h = 5$	$d = 2.5, h = 4$
Top	$5 \times d$	10	12.5
Back	$5 \times h$	25	20
Sides	$d \times h \times 2$	20	20
		Total: 55	Total: 52.5

Note: <u>All</u> possible choices for height and depth will produce less than 100 feet of area to be painted.

Perimeter of top $= w + w + d + d$ or $2w + 2d$
Width is 5.
For $d = 2$ ft, 14 ft of trim is needed.
For $d = 2.5$ ft, 15 ft of trim is needed.

Left panel

Name ______________________

UNIT 8 • TASK A

Woodworking

Materials: centimeter ruler

Mr. Leon makes wooden jewelry boxes. On the tops of the boxes he uses different kinds of wood to create geometric designs with triangles and quadrilaterals.

a. Draw a figure that is less than 15 centimeters by 15 centimeters. Use the figure to design a box top that includes triangles, rectangles, and parallelograms.

b. Use a centimeter ruler to measure the dimensions of the largest triangle and rectangle in your design. Write the dimensions on the diagram. Then find the areas of those figures.

c. Mr. Leon charges $0.25 per square centimeter to make his box tops. How much would he charge for the box you designed? $3.40

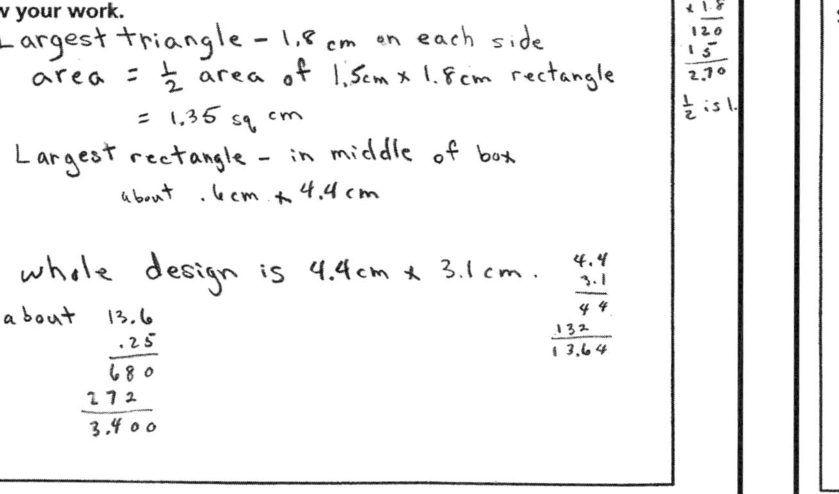

Show your work.

b. Largest triangle – 1.8 cm on each side

  area = $\frac{1}{2}$ area of 1.5cm × 1.8cm rectangle

      = 1.35 sq cm

  Largest rectangle – in middle of box

      about .6cm × 4.4 cm

c. whole design is 4.4cm × 3.1 cm.

  about 13.6
          .25
         ────
          680
          272
         ─────
         3,400

      1.6
    × 1.8
    ─────
      120
       15
    ─────
     2.70

    $\frac{1}{2}$ is 1.

      4.4
      3.1
     ────
       44
      132
     ─────
     13.64

Level 3 This work demonstrates the student's understanding of the task. Design is adequate. Dimensions are accurate. In part *c* calculations are accurate.

Performance Assessment   PA69

Right panel

Name ______________________

UNIT 8 • TASK A

Woodworking

Materials: centimeter ruler

Mr. Leon makes wooden jewelry boxes. On the tops of the boxes he uses different kinds of wood to create geometric designs with triangles and quadrilaterals.

a. Draw a figure that is less than 15 centimeters by 15 centimeters. Use the figure to design a box top that includes triangles, rectangles, and parallelograms.

b. Use a centimeter ruler to measure the dimensions of the largest triangle and rectangle in your design. Write the dimensions on the diagram. Then find the areas of those figures.

c. Mr. Leon charges $0.25 per square centimeter to make his box tops. How much would he charge for the box you designed?

Show your work.

b. Largest triangle 2cm × 2cm × 2cm

  Largest rectangle 4$\frac{1}{2}$cm × 1 cm

c. He would charge $4.63.

Level 2 This work displays partial understanding. The design is acceptable. Dimensions are basically accurate but areas are not calculated. Calculations for cost of tabletop should be $4.35.

Name ___________

Woodworking

Materials: centimeter ruler

Mr. Leon makes wooden jewelry boxes. On the tops of the boxes he uses different kinds of wood to create geometric designs with triangles and quadrilaterals.

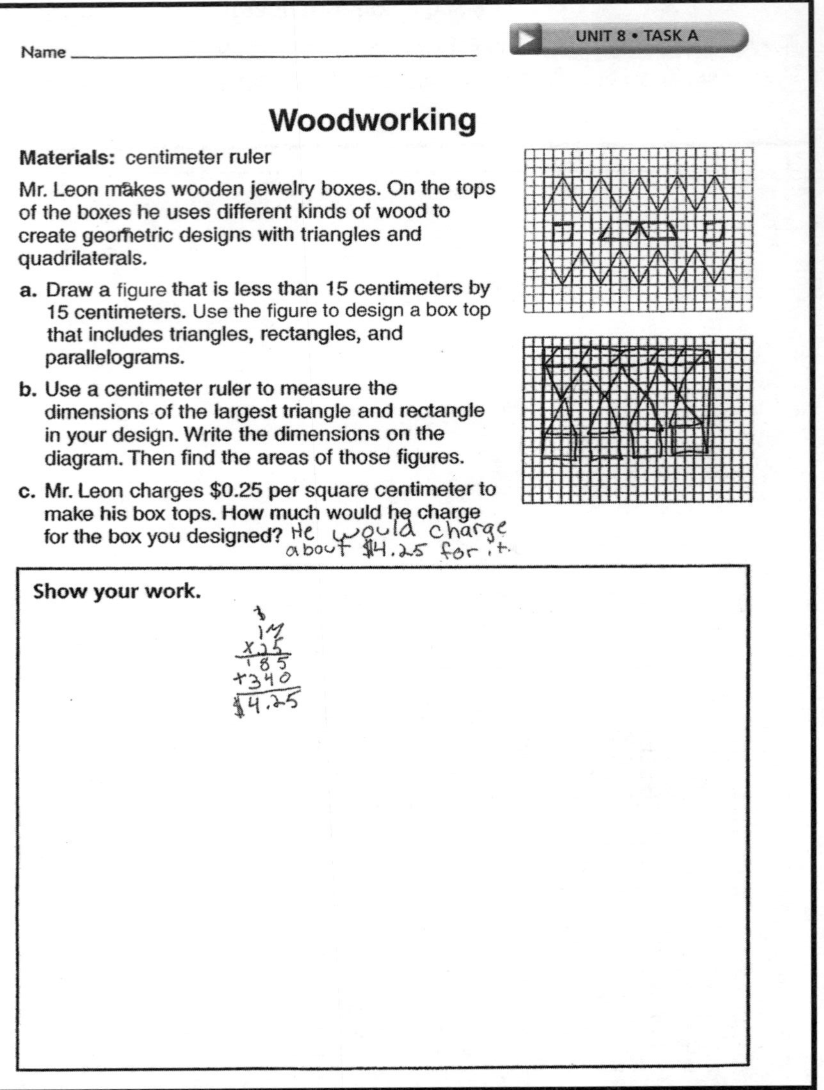

a. Draw a figure that is less than 15 centimeters by 15 centimeters. Use the figure to design a box top that includes triangles, rectangles, and parallelograms.

b. Use a centimeter ruler to measure the dimensions of the largest triangle and rectangle in your design. Write the dimensions on the diagram. Then find the areas of those figures.

c. Mr. Leon charges $0.25 per square centimeter to make his box tops. How much would he charge for the box you designed? He would charge about $4.25 for it.

Show your work.

$$\begin{array}{r} 1\,7 \\ \times 2\,5 \\ \hline 8\,5 \\ +3\,4\,0 \\ \hline \$4.25 \end{array}$$

Level 1   The student demonstrates little understanding of the task. Design is acceptable. Part *b* is omitted; the cost calculated in part *c* is incorrect based on the dimensions given in the diagram.

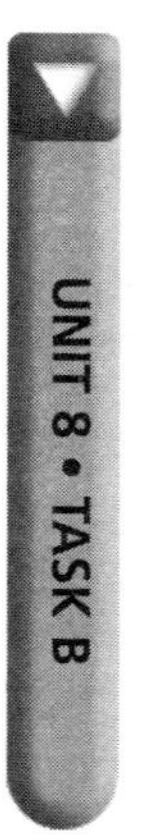

Level 3 paper

Name ______________

Storage Cabinet

Jason and his mother are designing a cabinet to store his sports equipment. They want the cabinet to have about 50 cubic feet of space.

a. Jason wants the cabinet to be 5 feet wide. Decide what dimensions for the depth and height of the cabinet you would use for a volume of about 50 cubic feet?

b. Jason wants to paint the cabinet. He reads on the paint can that the paint will cover 100 square feet. Does he have enough paint to cover the back, sides, and the top of the cabinet?

c. Suppose Jason decides to put wooden ledge around the top of the cabinet. How many feet of trim will he need?

Show your work.

a. It says about 50 cubic ft because the boards will take up some of the storage space. I would do width = 5 ft height = 5 ft depth = 2 ft

$25 \times 2 = 50$ cubic ft

b back – $5 \times 5 = 25$ sq ft
sides – $2 \times 5 = 10 \times 2 = 20$ sq ft
top – $2 \times 5 = 10$ sq ft

$25 + 20 + 10 = 55$ sq ft

Yes, he has enough paint.

c. $2 \times 5 + 2 \times 2 = 10 + 4 = 14$ ft of trim

Level 3   The student shows good understanding of the task. The answers are accurate and complete. The work is shown.

Level 2 paper

Name ______________

Storage Cabinet

Jason and his mother are designing a cabinet to store his sports equipment. They want the cabinet to have about 50 cubic feet of space.

a. Jason wants the cabinet to be 5 feet wide. Decide what dimensions for the depth and height of the cabinet you would use for a volume of about 50 cubic feet? Depth 5 ft width 5 ft hight 2 ft
50 cubic ft

b. Jason wants to paint the cabinet. He reads on the paint can that the paint will cover 100 square feet. Does he have enough paint to cover the back, sides, and the top of the cabinet? Yes it is enough

c. Suppose Jason decides to put wooden ledge around the top of the cabinet. How many feet of trim will he need? 20 ft

Show your work.

6.

$4 \times 5 = 20$ ft of trim

Level 2   This student has used appropriate dimensions. The student fails to give an explanation for part b, but correctly calculated perimeter for part c.

Name ___________________________

Storage Cabinet

Jason and his mother are designing a cabinet to store his sports equipment. They want the cabinet to have about 50 cubic feet of space.

a. Jason wants the cabinet to be 5 feet wide. Decide what dimensions for the depth and height of the cabinet you would use for a volume of about 50 cubic feet?

b. Jason wants to paint the cabinet. He reads on the paint can that the paint will cover 100 square feet. Does he have enough paint to cover the back, sides, and the top of the cabinet?

c. Suppose Jason decides to put wooden ledge around the top of the cabinet. How many feet of trim will he need?

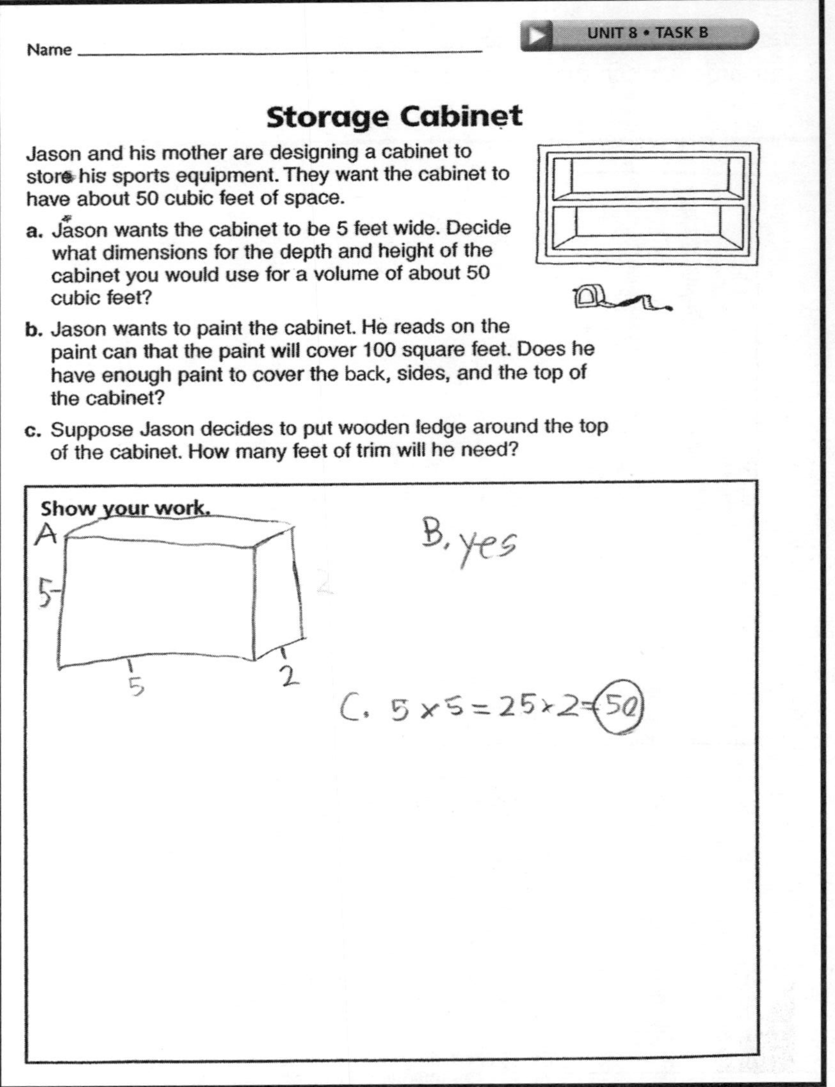

Level 1 This response shows partial understanding. The dimensions are appropriate. The response to part *b* is correct but no explanation is given. Student calculated area, not perimeter.

TASK A

Sale Time

Purpose
To assess student understanding of percent

Grouping
Individuals or partners

Time
10–15 minutes

Introduce the Task
Students complete a table by computing the discount and the sale price for items when they are given the regular price and the percent off for the items. Students then suggest items that could be purchased for a given amount of money.

TASK B

Probability Spins

Purpose
To assess student understanding of probability and equally likely outcomes

Grouping
Individuals or partners

Time
20–30 minutes

Introduce the Task
Students solve a problem of designing a spinner and listing possible outcomes. Students conduct an experiment and record results in a table. Students also determine probability of specific outcomes using a spinner.

Name _______________________________________ Date _______________________________________

TASK A

Sale Time

Performance Indicators	Observations and Rubric Score
_________ Makes a diagram or an organized list of possible choices. _________ Determines how number of choices changes when another item is included. _________ Suggests possible items for a given number of choices. _________ Shows work and explains how the answers were determined.	3 2 1 0

TASK B

Probability Spins

Performance Indicators	Observations and Rubric Score
_________ Designs a spinner using at least 2 different numbers or colors. _________ Lists possible outcomes of the spinner, determines and explains whether each outcome is equally likely. _________ Describes and conducts a probability experiment using the spinner. Records results in a table. _________ Finds the probability of the pointer landing on each number on Misha's spinner. _________ Determines and explains whether there is a greater probability of the pointer landing on an odd number or even number on Misha's spinner.	3 2 1 0

Total Score _________ /6

Sale Time

The Sports Shop is having a sale on soccer clothes and equipment.

a. Complete the table for the sale items.

Soccer Item	Regular Price	Percent Off	Discount	Sale Price
ball	$12.00	40%	$4.80	$7.20
shorts	$18.00	25%		
knee pads	$10.00	40%		
shirt	$15.00	25%		
shoes	$45.00	40%		

b. Justin has $50.00. He plans to spend 75% of his money on soccer sale items. Suggest some items he could buy. Tell how much he will spend and how much money he will have left.

____

____

____

____

c. On the last day of the sale, the three soccer balls that were left were sold for 60% off. The regular prices for the balls were $15.00, $22.00, and $30.00. Sara has $10.00. Which soccer balls can she buy? Explain how you know.

____

____

____

____

Probability Spins

Suppose you are designing a spinner like this to use in an experiment.

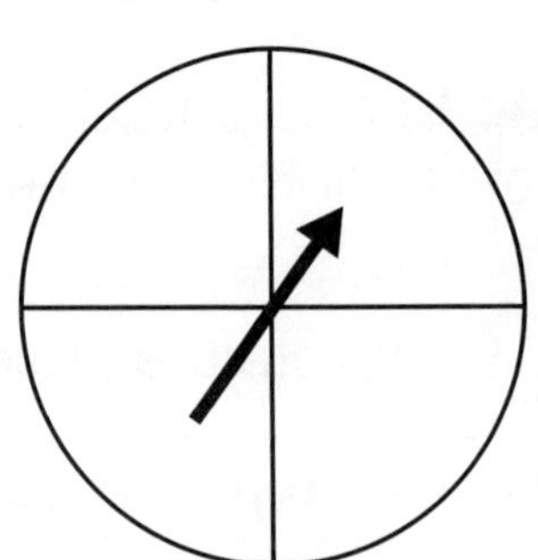

a. Use at least 2 different numbers or colors to design your spinner. List the possible outcomes of your spinner. Is each outcome equally likely? Explain.

b. Describe a probability experiment using your spinner. Conduct your experiment and record your results in a table.

c. Misha designed this spinner.
What is the probability of the pointer landing on each number on Misha's spinner? Which has a greater probability using Misha's spinner, the pointer landing on an odd number or on an even number? Explain.

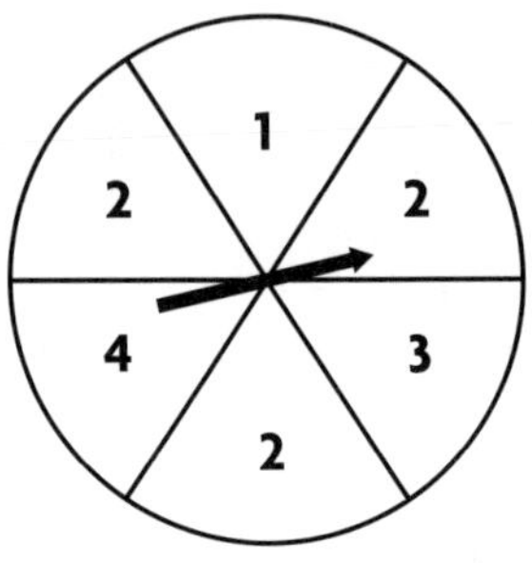

Show your work.

Name ______________________

Sale Time

The Sports Shop is having a sale on soccer clothes and equipment.

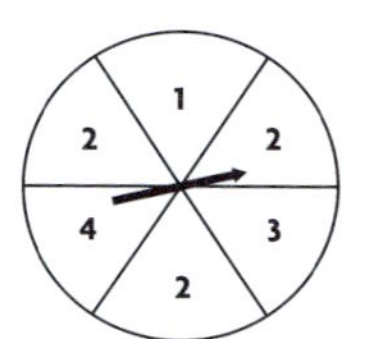

a. Complete the table for the sale items.

Soccer Item	Regular Price	Percent Off	Discount	Sale Price
ball	$12.00	40%	$4.80	$7.20
shorts	$18.00	25%	$4.50	$13.50
knee pads	$10.00	40%	$4.00	$6.00
shirt	$15.00	25%	$3.75	$11.25
shoes	$45.00	40%	$18.00	$27.00

b. Justin has $50.00. He plans to spend 75% of his money on soccer sale items. Suggest some items he could buy. Tell how much he will spend and how much money he will have left.

Answers will vary. Possible answer: shoes and knee pads; $33.00; $4.50

____

____

____

c. On the last day of the sale, the three soccer balls that were left were sold for 60% off. The regular prices for the balls were $15.00, $22.00, and $30.00. Sara has $10.00. Which soccer balls can she buy? Explain how you know.

the balls with regular prices of $15.00 and $22.00 because they will cost $6.00 and $8.80.

____

____

____

Performance Assessment   **PA75**

Name ______________________

Probability Spins

Suppose you are designing a spinner like this to use in an experiment.

a. Use at least 2 different numbers or colors to design your spinner. List the possible outcomes of your spinner. Is each outcome equally likely? Explain.

b. Describe a probability experiment using your spinner. Conduct your experiment and record your results in a table.

c. Misha designed this spinner. What is the probability of the pointer landing on each number on Misha's spinner? Which has a greater probability using Misha's spinner, the pointer landing on an odd number or on an even number? Explain.

Show your work.

a. Check students' work.

b. Answers will vary. A possible experiment for a spinner numbered 1–4 could be to predict how many times the spinner will land on 2 out of 30 tries. Check students' tables.

c. The probability of landing on 2 is 3 out of 6, $\frac{3}{6}$, or $\frac{1}{2}$; the probability of landing on 1, 3, or 4 is 1 out of 6, $\frac{1}{6}$.

PA76   Performance Assessment

Sale Time

The Sports Shop is having a sale on soccer clothes and equipment.

a. Copy and complete the table for the sale items.

Soccer Item	Regular Price	Percent Off	Discount	Sale Price
ball	$12.00	40%	$4.80	$7.20
shorts	$18.00	25%	4.50	13.5
knee pads	$10.00	40%	4.00	6
shirt	$15.00	25%	3.75	11.25
shoes	$45.00	40%	18.00	27

b. Justin has $50.00. He plans to spend 75% of his money on soccer sale items. Suggest some items he could buy. Tell how much he will spend and how much money he will have left.

He can spend $37.50.
If he buys the shoes, he can't buy anything else except maybe 1 thing. So I think he will buy shorts, knee pads, and a shirt. These cost $30.75. He has $6.75 left.

c. On the last day of the sale, the three soccer balls that were left were sold for 60% off. The regular prices for the balls were $15.00, $22.00, and $30.00. Sara has $10.00. Which soccer balls can she buy? Explain how you know.

She can buy the $15 ball because 60% off is under $10. She can't pay for the $30 ball because it would be $12. I think she might buy the other ball because it would cost $8.80.

Level 3 This student has completed all parts of the task adequately. The explanations include the necessary information.

Sale Time

The Sports Shop is having a sale on soccer clothes and equipment.

a. Copy and complete the table for the sale items.

Soccer Item	Regular Price	Percent Off	Discount	Sale Price
ball	$12.00	40%	$4.80	$7.20
shorts	$18.00	25%		
knee pads	$10.00	40%		
shirt	$15.00	25%		
shoes	$45.00	40%		

b. Justin has $50.00. He plans to spend 75% of his money on soccer sale items. Suggest some items he could buy. Tell how much he will spend and how much money he will have left.

Justin can buy shoes for $27. He can also get a shirt for $9. He has $14 left.

c. On the last day of the sale, the three soccer balls that were left were sold for 60% off. The regular prices for the balls were $15.00, $22.00, and $30.00. Sara has $10.00. Which soccer balls can she buy? Explain how you know.

The one for $15 and $22 will cost enough for $10. I know because 60% is more than half and it that is less than 10.

Level 2 The student did not complete the table. Answer for parts *b* and *c* are adequate.

Model Student Papers for
Sale Time

Sale Time

The Sports Shop is having a sale on soccer clothes and equipment.

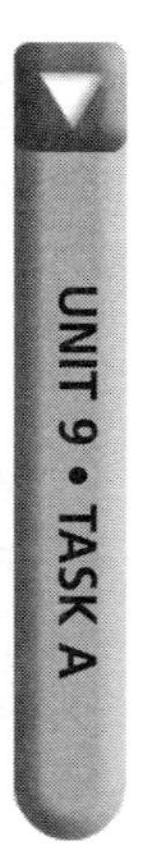

a. Copy and complete the table for the sale items.

Soccer Item	Regular Price	Percent Off	Discount	Sale Price
ball	$12.00	40%	$4.80	$7.20
shorts	$18.00	25%	25	17.15
knee pads	$10.00	40%	40	9.60
shirt	$15.00	25%	25	14.75
shoes	$45.00	40%	40	5

b. Justin has $50.00. He plans to spend 75% of his money on soccer sale items. Suggest some items he could buy. Tell how much he will spend and how much money he will have left.

He can spend 25 plus about 10 more. He can buy anything but the shoes.

c. On the last day of the sale, the three soccer balls that were left were sold for 60% off. The regular prices for the balls were $15.00, $22.00, and $30.00. Sara has $10.00. Which soccer balls can she buy? Explain how you know.

She don't have money

Level 1   The student shows no understanding of percent and discounts. Part *b* begins to explain how to find 75% of $50.

Left paper (Level 3)

Name ______________________

UNIT 9 • TASK B

Probability Spins

Suppose you are designing a spinner like this to use in an experiment.

a. Use at least 2 different numbers or colors to design your spinner. List the possible outcomes of your spinner. Is each outcome equally likely? Explain.

b. Describe a probability experiment using your spinner. Conduct your experiment and record your results in a table.

c. Misha designed this spinner. What is the probability of the pointer landing on each number on Misha's spinner? Which has a greater probability using Misha's spinner, the pointer landing on an odd number or on an even number? Explain.

Show your work.

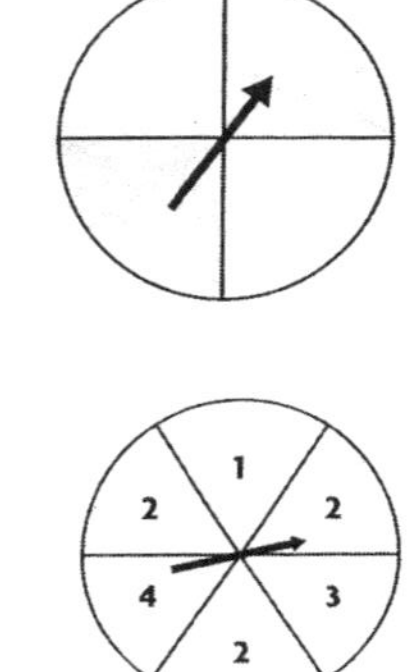

(b)

Outcomes	Spins
1	IIII
2	IIIIIII
3	IIIIIII
4	IIIIIIIII

(a.) [spinner: 1, 2, 3, 4]

Outcomes: 1 2 3 4

Each outcome is equaly likly, because each section is the same size.

(b.) How many times will I land on 1 out of 30 spins?

(c)
1-1 out of 6 chance
2-3 out of 6 chance
3-1 out of 6 chance
4-1 out of 6 chance

There is a greater chance of landing on an even since 4 out of 6 spaces have an even no.

Level 3 This paper shows a good understanding of the task. The student designed a spinner and listed all possible outcomes. The student conducted an experiment, recorded results, and explained questions.

Right paper (Level 2)

Name ______________________

UNIT 9 • TASK B

Probability Spins

Suppose you are designing a spinner like this to use in an experiment.

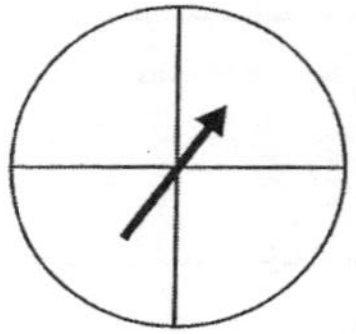

a. Use at least 2 different numbers or colors to design your spinner. List the possible outcomes of your spinner. Is each outcome equally likely? Explain.

b. Describe a probability experiment using your spinner. Conduct your experiment and record your results in a table.

c. Misha designed this spinner. What is the probability of the pointer landing on each number on Misha's spinner? Which has a greater probability using Misha's spinner, the pointer landing on an odd number or on an even number? Explain.

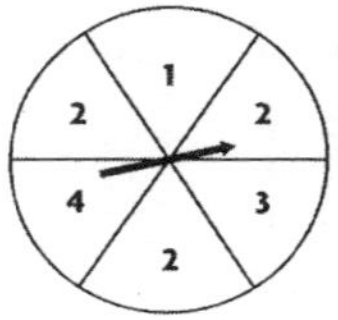

Show your work.

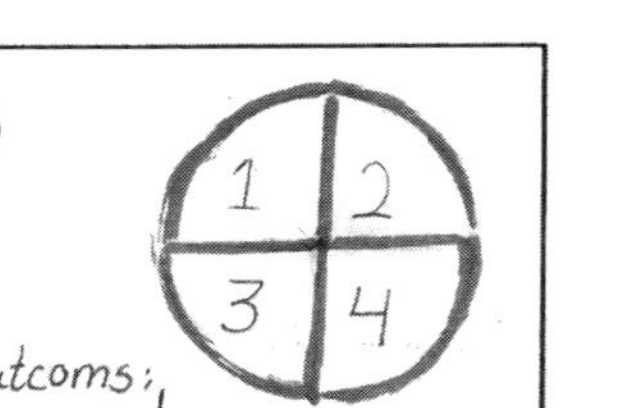

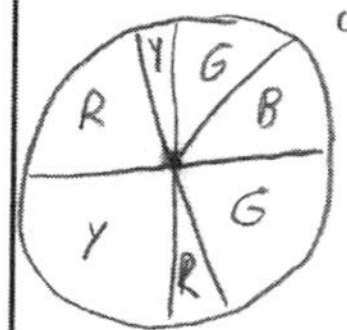

outcomes: Red Blue Green Yellow [spinner: Y G B G R R]

A) Each out come is not equally likely because, they some are small and some are big, or not same size

B) How many times will I land on Red out of 30 spins?

c) $\frac{1}{6}$, even because there are more even numbers than there are odd numbers

1-$\frac{1}{6}$ 4-$\frac{1}{6}$
2-$\frac{3}{6}$
3-$\frac{1}{6}$

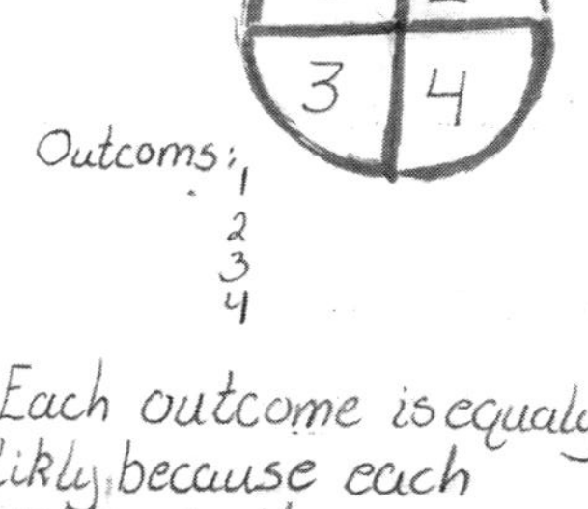

Outcomes	Spins
Red	THL
Yellow	THL I
Green	THL IIII
Blue	THL I

Level 2 All parts of this task were completed, but some are incorrect. The results recorded in the table reflect 26 spins, not 30. The explanation in part c is unclear.

Model Student Papers for
Probability Spins

Name ______________

Probability Spins

Suppose you are designing a spinner like this to use in an experiment.

a. Use at least 2 different numbers or colors to design your spinner. List the possible outcomes of your spinner. Is each outcome equally likely? Explain.

b. Describe a probability experiment using your spinner. Conduct your experiment and record your results in a table.

c. Misha designed this spinner.
What is the probability of the pointer landing on each number on Misha's spinner? Which has a greater probability using Misha's spinner, the pointer landing on an odd number or on an even number? Explain.

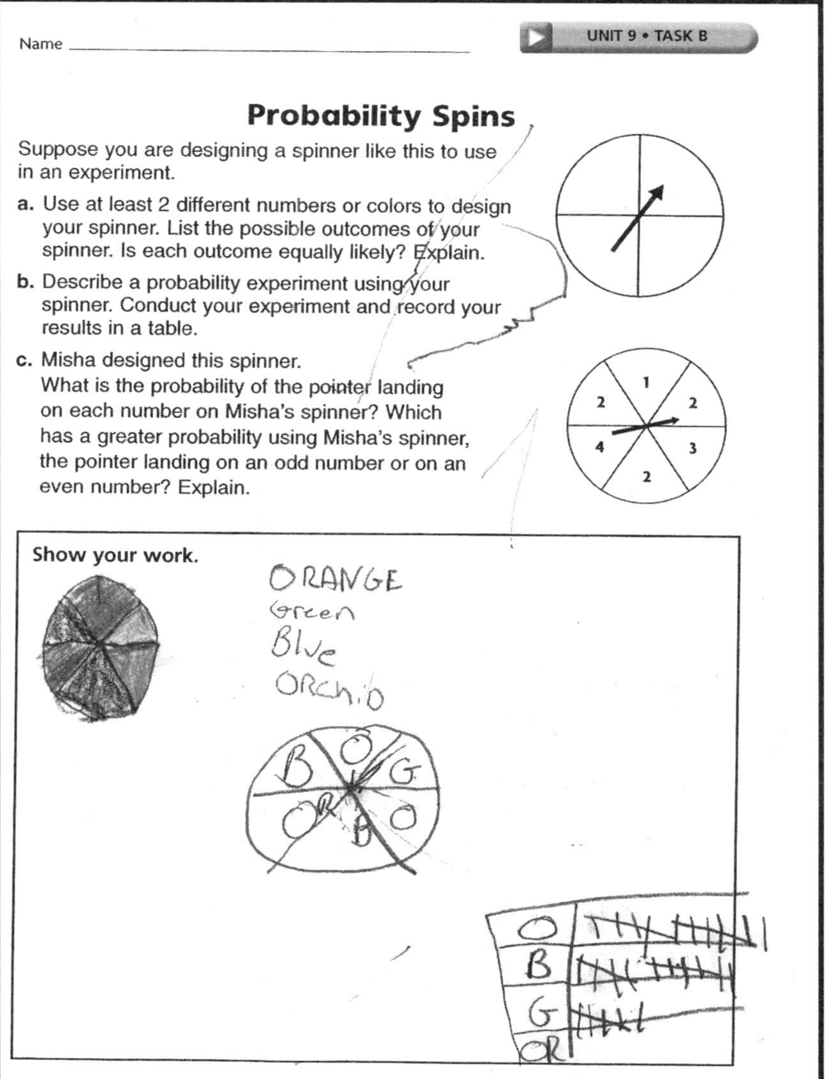

Show your work.

Level 1   This response is incomplete. The student designed a spinner and recorded data. The student did not describe the experiment or give the possible outcomes on their spinner.